A Survey of Entomology, Second Edition

Also by Gene Kritsky

The Quest for the Perfect Hive
Periodical Cicadas: the Plague and the Puzzle
Instant Origin
Insect Mythology (with Ron Cherry)
In Ohio's Backyard: Periodical Cicadas

A Survey of Entomology, Second Edition

Gene Kritsky and Frank N. Young, Jr.

A SURVEY OF ENTOMOLOGY, SECOND ED

Copyright © 2011 by Gene Kritsky and Frank N. Young, Jr.

All rights reserved. No part of this book may be used or re   by
any means, graphic, electronic, or mechanical, including pl   ng,
recording, taping or by any information storage retrieval sy
without the written permission of the author except in the
brief quotations embodied in critical articles and reviews.

iUniverse books may be ordered through booksellers or by   g:

iUniverse
1663 Liberty Drive
Bloomington, IN 47403
www.iuniverse.com
844-349-9409

Because of the dynamic nature of the Internet, any web ad
links contained in this book may have changed since publi
may no longer be valid. The views expressed in this work a   hose
of the author and do not necessarily reflect the views of th   r,
and the publisher hereby disclaims any responsibility for th

Any people depicted in stock imagery provided by Thinkst   dels, and such
images are being used for illustrative purposes only.
Certain stock imagery © Thinkstock.

ISBN: 978-1-4620-4449-8 (sc)

Library of Congress Control Number: 2011913711

Print information available on the last page.

iUniverse rev. date: 08/14/2020

TABLE OF CONTENTS

INTRODUCTION ... ix

CHAPTER 1: THE IMPORTANCE OF INSECTS ... 1
Agricultural Pests ... 1
Biological Control ... 2
Vectors of Disease ... 3
Insects' Influence on Culture ... 4
Insects as Pollinators of Plants ... 4
Industrial Materials Produced by Insects ... 6
Insects as Food ... 7
Honey, Honey Gathering, and Beekeeping ... 8
Insects as Tools in Scientific Research ... 10
Conclusions ... 10

CHAPTER 2: THE BASIC STRUCTURE OF INSECTS ... 12
Integument ... 12
Processes of the Body Wall ... 15
The Body Form of Insects ... 17
The Insect Body Plan ... 18
The Insect Head ... 19
The Thorax ... 24
The Legs ... 24
The Wings ... 25
The Venation of the Wings ... 28
The Abdomen ... 29
The Alimentary Canal ... 30
The Tracheal System ... 30
The Insect Circulatory System ... 31
The Excretory System ... 33

CHAPTER 3: MUSCLES, LOCOMOTION, AND FLIGHT ... 34
Locomotion ... 35
Functioning of the Leg Muscles ... 35
The Origin of Wings ... 37
Flight Muscles ... 39

CHAPTER 4: THE NERVOUS SYSTEM ... 43
Function of Nerves ... 44
Sense Organs ... 47
Organs of Vision ... 49

CHAPTER 5: INSECT REPRODUCTIVE SYSTEMS AND DEVELOPMENT ... 53
Metamorphosis ... 57

Insect Hormones and Molting	*60*
Pheromones	*62*

CHAPTER 6: INSECT BEHAVIOR — 64
Defensive Behavior — *66*
Reproductive Behavior — *67*
Communication in Insects — *69*

CHAPTER 7: INSECT EVOLUTION — 73
The Early Evolution of Insects — *77*

CHAPTER 8: APPLIED INSECT ECOLOGY — 81
What is an insect pest? — *81*
Integrated Pest Management (IPM) — *82*
Life Tables in IPM — *86*

CHAPTER 9: INSECTS' PLACE IN NATURE — 91
Reproductive Compatibility — *92*
What is a Species? — *93*
Higher Categories — *93*
The Genus — *94*
The Family — *94*
The Order, Class, Phylum, and Domain — *94*
Distribution of Insects — *95*
Sizes of Insects and Their Relatives — *96*

CHAPTER 10: ENTOGNATHAN HEXAPODS — 104
Class Protura — *104*
Class Collembola — *105*
Class Diplura — *106*

CHAPTER 11: THE PRIMITIVELY WINGLESS INSECTS — 108
Order Archaeognatha — *108*
Order Thysanura — *109*

CHAPTER 12: THE PRIMITIVELY WINGED INSECTS — 111
Order Ephemeroptera — *111*
Order Odonata — *114*

CHAPTER 13: THE POLYNEOPTERA — 117
Order Plecoptera — *117*
Order Embiidina — *119*
Order Phasmida — *121*
Order Orthoptera — *121*
Order Mantophasmatodea — *124*
Order Zoraptera — *124*
Order Blattaria — *124*

Order Isoptera	*126*
Order Mantodea	*129*
Order Dermaptera	*131*
Order Grylloblattodea	*132*

CHAPTER 14: THE PARANEOPTERA — 134
Order Psocodea — *134*
Order Thysanoptera — *136*
Order Hemiptera — *139*

CHAPTER 15: THE ENDOPTERYGOTA — 145
The Neuropterida — *145*
Order Megaloptera — *145*
Order Raphidioptera — *146*
Order Neuroptera — *147*
Order Strepsiptera — *158*

CHAPTER 16: THE HYMENOPTERIDA — 160
Order Hymenoptera — *160*

CHAPTER 17: THE PANORPIDA — 164
Order Mecoptera — *164*
Order Siphonaptera — *165*
Order Diptera — *166*
Order Trichoptera — *170*
Order Lepidoptera — *170*

APPENDIX 1: METHODS OF COLLECTING AND PRESERVING ARTHROPODS — 175

APPENDIX 2: IDENTIFYING THE ORDERS OF INSECTS AND NONINSECTAN HEXAPODS — 189

ADDITIONAL READINGS — 202

ABOUT THE AUTHORS — 203

GLOSSARY — 204

INDEX — 221

INTRODUCTION

I first met Frank N. Young, Jr. in 1972 while I was an undergraduate at Indiana University. It was the Viet Nam war era; I had long hair, and he was in the Army Reserve. While I was skeptical of this professor, I quickly discovered that he was a man in love with his occupation, and also clearly the most fulfilled man I had ever met. I eventually became his undergraduate teaching intern, and attended the University of Illinois to get my doctorate in entomology.

Frank and I stayed in close contact over the years. We shared a love of beetles and also wrote many papers together on the distribution of periodical cicadas in Indiana. In 1995, as his health declined, Frank asked if I could help him publish his last project: a textbook on entomology. Frank had taught many entomology classes for undergraduates, and he felt, as I do, that the available textbooks on the subject were needlessly cumbersome, geared as they were toward graduate students in entomology programs. Unable to find an appropriate text, he wrote his own chapters on insect biology and placed them on reserve in the Indiana University Undergraduate Library for his students' use. Included in these chapters were wonderful illustrations, concise descriptions of the orders, and carefully outlined methods of collecting and curation. Frank constantly revised and updated his work over the course of 20 years of teaching, continuing his research until his death in 1998.

Thanks to Frank's foresight, his work did not end with his passing. Very much as the surviving Beatles crafted "Free as a Bird" from John Lennon's demo tapes, my wife, Jessee Smith, and I refined his outlines and rough-draft chapters into a finished textbook. I updated his materials as needed, adding recently gained knowledge to Frank's chapters on insect biology and revising the chapters on taxonomy to reflect the changes made in that field in the light of molecular research. As some of Frank's line art was yet unfinished, Jessee applied her talents to transform his pencil drawings into finished illustrations and to add original figures w... were necessary. The manuscript was then "field-tested" in several... n undergraduate entomology classes and further revised in response... dents' comments and criticism.

The resulting book is the realization of... ultimate goal: to create a clear, concise textbook that will intr... udents from all backgrounds to insects and their biology, without... ss inundation of details relevant only to the graduate student. My h... at this book will act as a gateway, inspiring interest in these fasci... imals, providing students with an essential understanding of ento... and encouraging further learning in this field.

There are many who deserve acknowledgm... eir help. First, I wish to thank Jessee Smith, whose illustrations, ca... ing, and valuable suggestions greatly improved the final text. I al... my entomology students at the College of Mount St. Joseph and in... iculture program at the University of Cincinnati, who made hundred... ments that have helped to shape the text to the needs of its inte... ience. Finally, I thank Frank's wife, Francis Young, for her entl... support for this project.

This book is dedicated to all the students... rank and I have had in our entomology classes, and to all those wh... this book in the future.

Gene Kritsky

CHAPTER 1: THE IMPORTANCE OF INSECTS

Insects have been our constant companions since time immemorial. The first primates probably inherited sucking lice (Psocodea) from their insectivore ancestors, who may in turn have received them from their reptilian ancestors in the Mesozoic. When the first primates began building leafy nests in the trees, they also began an association with commensal cockroaches (Blattaria), silverfish (Thysanura), carpet beetles (Coleoptera), and clothes moths (Lepidoptera), whose counterparts still infest our homes. When the first humans began living in caves, they exposed themselves to infestation with the predatory bed bugs (Hemiptera), which had long been associated with the bats. As humans began to raise crops, store up food against the lean periods, and keep livestock, they began an association with a host of insects competing for crops, food stores, and even for the blood and flesh of domesticated animals. Among this last group of insects are those that are vectors or transmitters of human and animal diseases, and as the human population increased and people gathered together in villages, towns, cities, and great megalopolises, the problems caused by these insect associations increased exponentially.

Agricultural Pests

The realization of the importance of insects competing for crops came early. It was hard to ignore the migratory locusts (Orthoptera), to which there are numerous references in the Old Testament of the Bible. Even earlier evidence of our consciousness of the importance of insects is found in cave paintings, in figures on pottery, and in statues and carvings. In the United States, the Pilgrims in New England observed the emergence of the seventeen-year cicadas (Hemiptera). The early settlers in Kansas suffered from swarms of locusts that came off the prairies vacated by the bison, and the Mormons in Utah were saved from hordes of Mormon crickets (Orthoptera) only by the counter-invasion of seagulls.

As humans acquired an understanding of the importance of insect pests and agricultural enemies, we began to fight back. At first it was an unequal battle in which we could pit only the simplest tools—shovels, brooms, ditches, and fire—against the hordes of locusts or the masses of caterpillars. Gradually, however, more sophisticated weapons were introduced: soap, nicotine from tobacco, pyrethrum from the pretty flower of the same name, kerosene, and arsenicals such as Paris Green. "Entomology" as the science of the control of insects affecting agriculture began in the United States in the 1850's and grew along with the Land Grant Colleges and Experiment Stations, so that agricultural entomology in the United States is now more highly developed than in any other country.

In the early twentieth century, new and more efficient insecticides were discovered and introduced on a massive scale after World War II. DDT and other chlorinated hydrocarbons and the organic phosphorus compounds developed later seemed to be swinging the battle against the insects in our favor. Unfortunately, even though we have soaked the earth in insecticides to the extent that DDT was found in the flesh of penguins in the Antarctic, the insects have not given up. Today we find insecticide-resistant strains of flies, mosquitoes, fleas, cotton boll weevils, scale insects, grain moths, cockroaches, and almost any insect pest one can name. Pesticide-resistant cockroaches go merrily to sea in our submarines, and one could bet that uninvited insects have gone with our astronauts into space.

Biological Control

In the latter part of the nineteenth century, some workers developed the idea of pitting the insect pests' own parasites and predators against them. Early successes in the citrus industry in California were spectacular, but great difficulties have since been encountered in extending biological control, except in some favorable regions such as Hawaii. Remarkable success has been achieved in controlling the screw-worm fly (Diptera) by the release of males sterilized by irradiation. The manipulation of genetically altered insects promises even greater successes in the future.

Today, the inequalities in the battle against insects are decreasing. New crops resistant to insects are being produced and grown. New, more efficient, and safer insecticides are constantly being introduced. New ways of manipulating the biology of insects are being developed to reduce losses. Still, the losses are spectacular. In agriculture and stored products, insects probably cause in excess of five billion of dollars of loss each year in the United States alone. The loss of human life and productivity throughout the world due to insect carriers of disease is still beyond estimation.

Vectors of Disease

The recognition of the importance of insects as carriers of human disease came much later than the recognition of their importance to agriculture. For a long time, humans tried to ignore these tiny and apparently insignificant creatures. In the Middle Ages, it was part of the training of a lady not to scratch in public even if a flea, louse, or fly had bitten her.

In the sixth century C.E., the "black death," a complex of bubonic, pneumonic, and septicemic plague caused by a bacterium, *Pasturella pestis*, reached Europe and caused untold deaths before it subsided in the eighth century. In the fourthteenth century, a second great pandemic spread over Europe. This time it reached as far as England, and we have the literary accounts of Pepys and DeFoe to tell us of its horrors. It was not, however, until the third great pandemic began in the late nineteenth century that a real understanding was achieved of the cause of the disease and its relationship to the rat and the rat flea, *Xenopsylla cheopis* (Siphonaptera). As Hirst wrote: "For ages proud man, that seeming microcosm of universe, imagined that the great pandemics must be due to some grand cosmic cause, commensurate with its effect upon the human race. It was indeed chastening for the lord of the earth to learn that he must seek the cause of those terrible misfortunes in the patient study of the humblest of God's creatures, the vermin of his household."

The slave trade brought a tiny mosquito, *Aedes aegypti* (Diptera), out of the jungles of Africa and into the New World. Here this small creature, breeding in the water barrels, discarded pots, and broken bottles of trash heaps, carried the virus of the dread "Yellow Jack" or yellow fever, which fulminated into great epidemics. It was not until the work of Walter Reed, James Carroll, Jesse Lazear, and Aristides Agramonte in Cuba in 1900 proved the connection between the disease and this mosquito that the epidemics were brought under control. Dengue fever, a disease related to yellow fever, and carried by the same *Aedes aegypti*, still produces epidemics in warmer climates.

In World War I, thousands, perhaps millions, died in the trenches and beleaguered cities of Europe from typhoid fever, a disease caused by a rickettsia parasite and carried by the common body louse (Psocodea). In World War II, the Japanese may have lost Guadalcanal not so much because of the devastating attacks of American marines, but because they were unable to control malaria, a protozoan parasite transmitted by the *Anopheles* mosquito (Diptera). Large areas of Africa are still precariously occupied because of the danger to humans and animals of sleeping sickness, caused by trypanosome protozoa, which are carried by Tsetse flies (Diptera).

In a relatively short time, however, there have been great advances in the control of insect vectors and the diseases they transmit. Today, the rat flea can be easily controlled by dusting the burrows of the rat hosts with various insecticides. *Aedes aegypti* can be readily, if not easily, controlled by cleaning up the trash heaps. Body lice succumb to soap and water and insecticides, although in Korea they proved highly resistant to DDT. Global eradication of malaria through control of the *Anopheles* vectors is a real possibility. Unfortunately, as control methods, vaccines, and treatments are developed for old diseases, new ones emerge. Viral encephalitis transmitted from birds to humans by *Culex* mosquitoes (Diptera) is now a serious problem in the United States. In war-disturbed areas, in disasters, and in underdeveloped regions, even the well-known insect-borne diseases are more than holding their own.

Insects' Influence on Culture

Our account with the insects is not entirely on the debit side, however. Insects have impeded human progress in some areas, but they have also aided it in others. We can only guess about the effects that insects may have had upon ancient humans in the development of language, music, and poetry. In the Far East, crickets (Orthoptera) are still highly prized as "singers," and there is an active market for them, with some desirable species being transported long distances or specially bred. The "songs" of the cicadas (Hemiptera) are admired in Japan and imitated in music and poetry. In Western culture, the influence of insects has even been noted in classical music such as Rimsky-Korsakoff's "The Flight of the Bumblebee."

Insects may have had many other indirect effects upon human culture. The perfection of the clay nests of the mud-dauber wasps (Hymenoptera) could well have inspired the first pottery. Who can say that the tiny nets of the hydropsychid caddisflies (Trichoptera) did not first suggest the fishing net? More developed cultures saw insects as examples for human activities or used them to inculcate moral and ethical values, as indicated in axioms such as Proverbs 6:6: "Go to the ant, thou sluggard; consider her ways, and be wise."

Insects as Pollinators of Plants

One of the most profound effects of insects is a result of their relationship with the flowering plants. Most of the colors of flowers have evolved to attract insects to them to assure pollination, and the evolution of the higher plants and the insects is very closely related. Some of the coadaptations to assure cross-pollination are remarkable. For example, some of the orchids pollinated by Scoliid wasps (Hymenoptera) have developed flowers with pendant petals closely mimicking the markings on the abdomen of the female wasps. Males are attracted to these flowers and attempt to copulate with them (in some cases actually depositing sperm upon the flower) and, in the process of moving from one flower to another, they cross-pollinate them. The Yucca moth, *Tegticula alba* (Lepidoptera), is equipped with special maxillary tentacles with which the female collects the pollen of the yucca flowers and places it upon the stigmas after depositing her eggs in the ovary. Since the yucca flower is not suited for self-pollination, this mutualism assures the reproduction of the plant, and since large numbers of seeds are produced, the moth larvae are assured of sufficient food to complete their development.

Insect pollination is of great importance in agriculture. Many crops, such as wheat and other cereals, are wind-pollinated, but others, particularly orchard crops, set only small numbers of seed or fruit without being visited by insects. Cucurbits such as melons, squash, cucumbers, and pumpkins are largely dependent upon bees (Hymenoptera) for cross-pollination, and commercial production is difficult and uncertain when suitable pollinators are absent. In Mexico and other areas, production of certain cucurbits is dependent upon specific species of bees, which visit the flowers only during brief and specific times of the day.

The fig wasps (Hymenoptera) are particularly interesting pollinators. At least some figs, if not all, cannot be fertilized by any other insects. The small wasps penetrate into the peculiar flowers to lay eggs and, in doing so, introduce pollen from male flowers that they have previously visited or in which they were bred. In the Smyrna fig wasp, *Blastophaga psenes*, the males are wingless and never leave the male flowers in which they breed. The female wasps must visit them in order to be fertilized. Since the male flowers are borne on separate trees from the female flowers from which the fruit develops, both kinds of trees must be grown to assure pollination. From very ancient times, growers of these figs have made a practice of taking branches from the male trees, or caprifigs, and hanging them in the fruit-bearing trees. Whether pollination is actually necessary for fruit development is, however, uncertain, as some figs definitely produce fruit in the absence of male trees or wasps.

Red clover, which is an important forage plant, must be pollinated in order to bear seeds. Honey bees and bumble bees (Hymenoptera) are particularly important in this regard. When red clover was first introduced into New Zealand, it produced very poor crops of seeds until bumble bees were introduced. Today, some areas of New Zealand have the highest yield of clover seed of any part of the world.

Red clover pollination is also the basis of an anecdote attributed to Charles Darwin. England's predominance as a beef-producing country, and in consequence its economic prosperity, as Darwin said, was due to its large number of widows and spinsters. Widows and spinsters keep cats. Cats catch mice. Mice destroy bumblebee nests for the honey and larvae. Therefore, the more widows and spinsters there are, the more bumblebees there are to pollinate the clover, and this leads to la[...] lover fields upon which the famed English beef is fattened.

Industrial Materials Produce[d] [I]nsects

In the past, insects were of great importa[nt ...] ducing materials used in industry. These include important produc[ts ...]d today, such as silk, wax, shellac, and lacquer. The silkworm, *Bo[mbyx ...]* (Lepidoptera), is today a domesticated animal that forms the basis o[f ...] ndustry in China and Japan and some parts of Europe. The larvae [... mo]th are grown on the leaves of the white mulberry tree, and the silk i[s ...]ed by unwinding the cocoons that the larvae spin when they are read[y ...]te and transform into the adult moth. Despite the development of m[any ...]etic fibers, silk is still highly prized for its beauty and other properties[.]

Insect waxes still in use are primarily those [of ...] bees, but in the past other insect waxes were also important. *Ce[reus ...]riferus*, a lecaniid scale (Hemiptera: Coccidae) produces a white wax [... a]nd other species of *Ceroplastes* are used for wax production in other[... i]n China, a white wax produced by another lecaniid scale, *Ericerus pe[la]* [fo]rmerly used as a substitute for kerosene. Most of these waxes hav[e ...]een displaced by petroleum waxes, and even the honey bees are [...]n their work by partially prefabricated combs.

Lac, an important ingredient of shellacs, [... varnishes] and lacquers, is the natural product of another lecaniid scale, *La[...]*, which lives on various trees in India. The lac is the shell-like scale [produce]d by the insect as a protective covering. The scales are scraped off th[e ...]nd rendered into stick lac, which is still imported into the United St[ates ... var]ious plastics and synthetics now substitute for lac, although there a[re ...] special processes for which the natural product is considered superio[r.]

Dyestuffs from insects are now largely re[placed by ...]aniline, or coal tar, dyes. The Greeks and Romans, however, use[d ... dy]es from various species of *Kermes* (Hemiptera: Coccidae), which f[eed on ...] species of oak, *Quercus coccifera*. In northern Europe, a red [... was] obtained from *Porphyrophora polonica*, another scale that lives on t[he ... roots] of a smartweed, *Polygonium cocciferum*. Both of these dyes were re[placed] commercially by cochineal, which is produced by a scale, *Dactylopius [...* It] came originally from Mexico. This scale feeds on the Nopal cactu[s ... *Opunti*]*a coccinellifera*, and other cacti. It was introduced in several places a[round th]e Mediterranean region and did particularly well in the Canary Islan[ds wher]e it once formed the basis of a considerable industry. Cochineal is s[till used] as a red coloring in commercially produced foods, cosmetics, and beve[rages.]

Insects also produce a variety of miscellan[eous sub]stances that have been used in the past. Axin, a substance used in m[edical pr]eparations and as a varnish, comes from a Mexican scale, *Llaveia axin[...* G]alls of gall wasps

(Hymenoptera) were once used as sources of tannic acid for the manufacture of ink. Cantharidin, made from a dried beetle, the Spanish fly (*Lytta vesicatoria*), is still in use, particularly in animal husbandry. Surgical maggots, the sterile larvae of various blow flies or flesh flies (Diptera), were once important in medicine, particularly in brain surgery and the treatment of bone diseases. The use of these maggots is making a comeback. The venom of bees and wasps is also used in the treatment of arthritis and rheumatism, but its use is still experimental.

Insects as Food

Insects are of immense value as food for fish, birds, and other animals directly useful to us, but we can hardly estimate the importance of insects in the diet of early humans. Many hunter-gatherers living today practice entomophagy, and in some cases, insects form a significant part of the diet during certain periods of the year. Insects are a rich source of vitamins, fats, and proteins, and thus make excellent supplements for diets that are high in carbohydrates. Curiously, we are reintroducing insects as novelty items in our diet; for example, fried maguey worms (Lepidoptera) from Mexico, roasted locusts, chocolate-covered ants (Hymenoptera), sugared bee larvae, and preserved silkworm pupae from Japan.

In ancient times, insects were eaten freely in almost every region of the world. The "manna" eaten by the Children of Israel during their wanderings in the deserts of Sinai was apparently the honeydew secreted by a scale insect, *Trabutina mannipara*, which lives on tamarisk trees in many places around the Mediterranean. This material was called "manna" by the Arabs and used by them as food even in the late nineteenth century. The honeydew closely fits the descriptions of the "manna" in the older versions, but later priestly interpretations seem to have obscured the issue.

Migratory locusts were not only serious enemies of crops, but were also an important food supplement in many regions of the world. They are still eaten in areas of the Middle East, Asia, and Africa. In Okinawa during World War II, the second author personally obtained specimens of large grasshoppers, which he could not catch, from small boys who were catching them for food. In the United States, the native people of the Plains and even the early settlers in Kansas ate locusts during times of famine. Cicadas, particularly the nymphs and newly emerged adults, are used for food in southeastern Asia, and were once eaten in the United States by the Native Americans.

Except among a few cultures, however, insects have become mainly condiments or luxury items. In China, water beetles (Coleoptera: Dytiscidae), water bugs (Hemiptera: Belostomatidae), and pentatomid bugs (Hemiptera: Pentatomidae) are still marketed and used for making sauces or eaten as salted snacks. Some astoundingly evil-smelling bugs (Hemiptera) are used as seasoning.

In Africa, termites (Isoptera), locusts, and many kinds of beetles and moth larvae are eaten. In some cases they are collected, prepared, and sold in the markets. In South America, many of the native groups eat termites, ants, and other insects. In some areas, the termites are gathered from the nests by using a long straw or slender twig, which is wetted in the mouth. The termites are swept off by running the straw back through the mouth. This habit may have been acquired by observing the large anteaters that use their long tongues in much the same manner.

In Mexico, the eggs of a corixid bug (Hemiptera: Corixidae) are gathered from the large lakes that form in the desert regions. These are prepared as cakes and sold in many markets. Children in Mexico and the southwestern United States still eat honeypot ants (fig. 1.1), which store honey by feeding it to selected members of the colony that become so replete that they cannot move, but hang like living honey sacs from the roof of the nest.

1.1 The honeypot ant is prized for its sweet, honey-filled abdomen.

Up to historic times, some of the native peoples of the Pacific Northwest were foragers who ate many insects. In some areas, they gathered the larvae of a butterfly, *Neophasia menapia*, which feeds upon pine. They built fires beneath heavily infested trees and collected the larvae as they dropped to the ground to avoid the smoke. The larvae were then dried and formed into cakes, which could be used to enrich stews or be eaten as such.

Honey, Honey Gathering, and Beekeeping

The insect product of greatest importance to humans is honey, produced by many species of wild and domesticated bees. Honey is now an epicurean delicacy, but for millennia, it was the only concentrated sweet known to us. Paleolithic cave drawings in Araña, Spain, depict with unmistakable clarity a man gathering honey on a rope ladder against a rock cliff (fig. 1.2). In one hand he holds a pot or bag for the honey and in the other a smoke smudge to protect him from the bees. The bees, stylized and greatly magnified, are shown attacking him.

1.2 Rock painting depicting honey-gathering, discovered in the Cuevas de la Araña near Bicorp in Valencia, Spain.

This drawing is particularly interesting because the techniques illustrated are remarkably similar to those still used by the Veddas in Sri Lanka to gather honey from rock-dwelling bees. The process used by the Veddas includes a complex ritual accompanying the activities and rules as to who can help. For example, the man or boy holding the rope ladder must be the climber's father-in-law or brother-in-law—the only members of the tribe not eligible to marry the climber's widow.

Greek warriors carried honey into battle with them—a ready reserve of quick energy for the moment of trial. In the *Anabasis*, Xenophon describes honey-seeking by the Greek mercenaries returning home after the death of Cyrus. His account of poisonous honey found in the mountains was long discounted because honey is usually the safest food in almost any form or condition. In more recent times, however, poisonous honey has been found to occur in the Appalachian mountain region, where it is associated with the honey made from blooms of *Rhododendron*.

Beekeeping was also a very early development in human culture. In the Old World, apiculture was centered on the honey bee, *Apis mellifera*. In the New World, before *Apis mellifera* was introduced from Europe, beekeeping utilized many kinds of stingless bees of the family Meliponidae. Among the Greeks and Romans, the art of beekeeping was well developed. They apparently acquired the techniques from the Egyptians, who made cylindrical horizontal hives out of clay and straw and stacked them like logs. With the Romans, beekeeping spread widely over Europe and remained there even after the retreat of the Roman legions.

One of the special qualities of honey is that it can ferment naturally, or be encouraged to do so, to form a heady, intoxicating drink called mead. The mead halls of Beowulf and his Anglo-Saxon descendants were centers of social life. There the minstrels sang and told their heroic tales and laid the foundation of the English language. Perhaps, without the bee and its honey, this tale might need to be told in some language developed under the influence of fermented turnip peelings. *Nyet?*

Insects as Tools in Scientific Research

Another contribution of insects to human progress has been their usefulness in scientific research. They are small, easily bred, and, having no SPCA for their protection, can be handled in ways impossible with mice, rabbits, guinea pigs, or dogs. The fruit flies, *Drosophila melanogaster* and related species, are the basis of much of our knowledge of genetics. Basic contributions to biology have also been made through research on the physiology, endocrinology, and other aspects of insects. Population ecology has been advanced through studies of the tiny flour beetle *Tribolium confusum* and its relatives. Insects are also valuable in the assay of drugs. Our attempts to control insects have paid special dividends through basic knowledge that has been found applicable in other fields.

Insects are also the focus of genome projects. The genome of the honey bee has been sequenced and is leading to new insights in honey bee physiology, behavior, and evolution. The genome of the mosquito that spreads malaria is also providing new clues for dealing with this disease, and the study of the DNA of lice is providing new information on the evolution of our species.

Conclusions

However we feel about insects—whether we consider them friends or foes—we must conclude that they have been and are still of great importance to us. We must continue to study them, not only to combat their damage, but also to discover and appreciate their usefulness. Entomology, the study of insects, is one of the major work areas of modern science, and the many special fields within entomology are making basic contributions to human knowledge, as well as to human welfare.

CHAPTER 2: THE BASIC STRUCTURE OF INSECTS

All insects have a basic structural pattern. This basic form, although it is elaborated in a great variety of ways, has remained rely constant from Paleozoic times down to the present. It is obviously the "body plans" of life that has proven highly successful. In fact, judged all criteria except individual size, the insect pattern has proven to be the adaptable of all patterns among multicellular animals.

Integument

The insect cuticle is a complex association of substances held in a crystalline lattice of chitin and protein. Chemically (fig. 2.1) is an amino sugar (n-acetyl-D-glucosamine) in which the molecules form long, unbranched chains. These chitin molecules are linked to protein matrix by hydrogen bonds between the chitin and protein molecules.

2.1 Chitin molecule

Physically, chitin is not hard and brittle, but very elastic and tough. In some membranes, it is associated with rubber-like proteins, known as resilins, which are almost perfectly elastic. Molecular changes to the protein matrix harden the cuticle, producing the tough parts of the exoskeleton.

The cuticle (fig. 2.2) is secreted by a layer of columnar or cuboidal epithelial cells, which begin their activities during the embryonic stages. These cells transfer materials from the blood and probably synthesize many of the specific compounds found in the cuticle. A section through the integument of an insect shows three distinct layers of non-living materials in the cuticle (the epicuticle, exocuticle, and endocuticle) overlying the epithelial layer of living cells. The epithelial cells also secrete an internal basement membrane, which probably functions selectively in the movement of materials from the blood into the integument.

Except in very thin cuticles, the layers are deposited through helical pore canals, which extend from the epithelial cells through the two inner layers. During the development of an insect, most of the materials in the endocuticle layer and much of the exocuticle layer are reabsorbed before the cuticle is shed, which must be done periodically to allow for further growth. Thus, only the epicuticle and part of the exocuticle are lost at each molt. In the Thysanura and Archaeognatha (primitive wingless insect orders), the cuticle is shed periodically throughout the life of the animal, but in the more advanced insects, there is a final definitive adult molt after which the insect never again sheds the cuticle.

The epicuticle (fig. 2.3), though only 0.1 to 0.3 microns thick, is chemically extremely complex. Its outer layer, secreted by the epithelial cells, is composed of waxes or lipids, which are the principal protection of terrestrial insects against desiccation. The surface of the wax layer is often further protected by a layer of cement or waxes that is secreted by specialized dermal glands.

Beneath the cement and wax layers are layers of polyphenols and cuticulin, and also a complex, non-chitinous layer of lipoproteins and other proteins. The waxes in the epicuticle are hard or semi-hard at normal temperatures, but they have a "break point" above which they quickly liquefy. Therefore, some insects kept in a temperature around 29.5° C lose the protection of the waxes and quickly die of desiccation, while other insects may have a transitional temperature 20° to 30° C or higher.

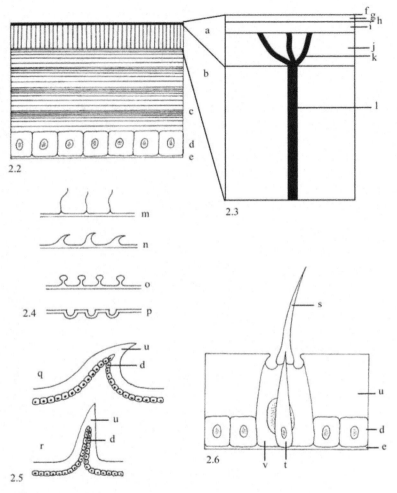

2.2 Insect integument.
2.3 Epicuticle and exocuticle.
2.4 Non-living epicuticle processes.
2.5 Cellular processes of the cuticle.
2.6 Seta.

a-epicuticle
b-exocuticle
c-endocuticle
d-epidermis
e-basement membrane
f-cement layer
g-wax layer
h-superficial layer
i-outer epicuticle
j-inner epicuticle
k-wax filament
l-pore canal
m-hairs
n-hooks
o-knobs
p-punctures
q-spine
r-spur
s-seta
t-trichogen cell
u-cuticle
v-tormogen cell

The epicuticle contains a number of enzymes to toughen and usually darken the cuticle in a process called tanning. Tanning and hardening are related but separate processes. Hardening results in a complex protein called sclerotin, and the hardening process is restricted to the specific areas of the cuticle that form sclerites. Between the sclerites, the integument has much the same chemical nature, but hardening does not occur, although special toughening and strengthening substances (resilins) are formed. Only the epicuticle and part of the exocuticle are hardened, or sclerotized, and the membranes between the sclerites remain soft and flexible. Darkening of the integument is known as melanization. When newly hatched or immediately after molting, insects are almost completely white or colorless. The hardening, tanning, and darkening processes usually begin immediately. The exocuticle and endocuticle are less complex than the epicuticle and contain chitin throughout.

While the cuticle provides muscle support and protection for the insect, it does have an important limitation in that it cannot grow. This requires insects to undergo the molting process, during which they shed the cuticle and produce a larger one in order to increase in size.

Molting is divided into two steps. Apolysis is the separation of the old cuticle from the epidermis, and ecdysis is the actual shedding of the old cuticle. Apolysis starts with the epidermal cells becoming elongated into a thicker layer of columnar cells. The old cuticle separates from the epidermal cells when they return to their normal cuboidal shape. The space between the old cuticle and the epidermis then fills with molting fluid, which is produced from the epidermal cells. Molting fluid contains enzymes that digest the old endocuticle.

The new cuticle starts to form in the space between the old cuticle and the epidermis with the formation of the procuticle, which is composed of both endocuticle and exocuticle. Molting fluid then digests the old endocuticle and its molecules are recycled into the new cuticle. As the old procuticle is broken down, it becomes thinner, while the newly forming procuticle increases in thickness. Eventually, only the epicuticle and a small portion of the procuticle are left, with the new epicuticle and procuticle inside. Finally, the remnants of the old cuticle are shed in the process of ecdysis.

When the insect is no longer restricted by the old cuticle, it can increase its size by expanding the newly formed cuticle. After expansion, the new cuticle can sclerotize and harden.

Processes of the Body Wall

A variety of processes extend from the body wall both externally and internally. They perform many different functions and may be roughly classified as either cellular or non-cellular processes. Non-cellular processes (fig. 2.4) are extensions of the epicuticle that are formed when the cuticle is laid down and are non-renewable until the cuticle is shed. They take the form of "hairs," hooks, tiny spines, bumps, knobs, tubercles, ridges, and other forms. Cellular processes (fig. 2.5) are extensions produced by epidermal cells, and they are modifiable or repairable as the insect grows.

The seta is an important type of cellular process (fig. 2.6). It is formed by a specialized epidermal cell called a trichogen cell, which extends through the cuticle and forms the base of the seta. The trichogen cell is usually associated with a tormagen cell that forms the socket in which the seta is placed. Setae are often hollow spines of hardened cuticle, but they may take on many forms. Types of seta range from simple hairs, plumose hairs, and flattened hairs to indescribably intricate forms found on the Collembola and Diplura (two noninsectan hexapods). Many setae are sensory and are associated with sensory cells and nerve endings, whereas others inject poisons when their tips are embedded and broken off in the tissue of another organism.

The flattened scales of Thysanura and the flat, striated scales of Lepidoptera (the moths and butterflies) are modified setae. Scales may have secondary non-cellular processes or ridges on their cuticle. In the Morpho butterflies, for example, the striations of the scale simulate an optical grid that absorbs light differentially so that only a single color is reflected at any given angle of incidence of the light. This produces marvelous iridescent metallic coloration that changes from purple to red to blue to green as the insect moves.

Multicellular processes of the body wall differ in being either fixed or articulated. Typical fixed processes are the spines (fig. 2.5) on the legs of grasshoppers and other insects. These outgrowths of the body wall are continuous with the cuticle and secreted by epidermal cells. Articulated appendages typically include legs, antennae, cerci, portions of the genitalia, and wings. In many insects, however, spurs are characteristic structures on the legs or other body parts. Spurs differ from spines mainly in that the cuticle at their base forms a membrane, allowing movement. The claws of insects are moveable spurs in which there are muscle attachments.

Internally projecting processes are called apodemes, and are largely associated with muscle attachments. The inner surface of the pleural suture (a groove on the side of the thorax), for example, is the external expression of a ridge-like apodeme. The head is supported internally and provided with muscle attachments by a complex of apodemes forming the tentorium (a scaffold inside the head). Similar apodemes in the thorax provide muscle attachments for the wing muscles.

The Body Form of Insects

The body form of arthropods is largely determined by the pattern of sclerites in the integument. The insect pattern of sclerite arrangement, although showing considerable variation in minor details, remains constant throughout the class.

Sclerites are hardened, definitively limited areas in the cuticle that meet along seams called sutures. Open sutures are flexible membranes between sclerites, and fused sutures are the lines of fusion of two or more sclerites. Sclerites take on many different forms; they may be flat, curved, cylindrical, forked, or elaborated in other ways. Their original function may largely have been protection, and they still play an important part in this role in insects such as the beetles, which are essentially flying tanks. Their most important role, however, is to serve as points of attachment for muscles.

The first sclerites established in the body wall of ancient insects were the dorsal and ventral plates of the body segments. These are respectively the terga (singular: tergum) and the sterna (singular: sternum) (fig. 2.7). The tergum is a saddle-like plate curved over the dorsal surface of a segment, and the sternum is a flatter plate beneath the segment. The sides, or pleura, of a segment are usually flexible membranes, which allow the dorsal tergum to overlap the ventral sternum when muscles running from one to the other contract.

The longitudinal contractions and extensions of the body are more complex, and two types of segmentation can be distinguished in living insects. Primary segmentation (fig. 2.8) is seen in insects that retain a worm-like form, such as the larvae of many insects. In these insects, the integument usually remains rather thin and sclerites are small and widely separated. The body segments are divided by intersegmental grooves or folds, which are simply circular constrictions of the integument. Secondary segmentation (fig. 2.9) involves the enlargement and overlapping of the sclerites of the body segments, thus changing the position of the intersegmental lines so that they obscure and protect the unsclerotized intersegmental grooves. In this manner, secondary segmentation allows contraction of the body so that the sclerites provide more effective "armor," but the contraction can be relaxed to allow for feeding, reproduction, breathing, and other functions.

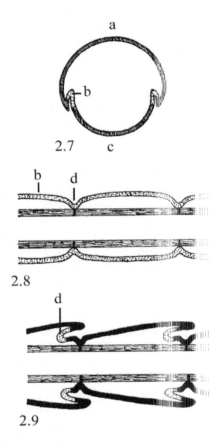

2.7 Cross-section of insect body.
2.8 Primary segmentation.
2.9 Secondary segmentation.
a- tergum
b- membrane
c- sternum
d- intersegmental groove

The Insect Body Plan

Adult insects (fig. 2.10) usually have a he： ⅼ s., and abdomen, three functional pairs of legs, and a single pair ınae. With the exception of primitive wingless insects, most adulⅰ ily recognized by the presence of wings or pleural sutures on the sid ⅼ thorax. Winged insects and secondarily wingless insects are biologic ⅰ ir in having some degree of change in appearance or metamorphosis between the immature and adult stages, and in not molting after sexual maturity is reached. All winged insects practice internal fertilization with some form of copulation.

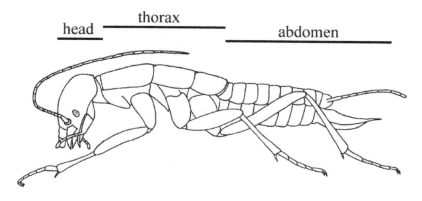

2.10 The insect body plan.

The Insect Head

The definitive insectan head (figs. 2.11, 2.12, and 2.13) is derived from six fused segments of the hypothetical insectan ancestor. Vestiges of sutures are recognizable on the posterior sides of the head in some Archaeognatha and Thysanura, but the dorsal sclerites of the head segments are fused to form a head capsule which encloses the brain and on which simple and compound eyes may be located. The antennae are segmented appendages articulating with the head capsule.

In the front, the head capsule is usually separated by a fused suture from the clypeus, to which a movable upper lip, the labrum, is articulated (fig. 2.11). Four segments are represented by the mouthparts, showing that they are clearly derived from segmented appendages. The hypopharynx, or tongue, consists of paired appendages that are fused together and usually contained within the mouth. In some insects, tiny vestigial feelers (superlinguae) are represented at the sides of the hypopharynx (fig. 2.21). The mandibles are the pair of appendages on either side of the mouth (figs. 2.11, 2.12). They usually have distinct biting surfaces or cutting edges, and are usually without indication of segmentation. They also have two articulating points, the condyles (fig. 2.14j), except in the order Archaeognatha, which have a single articulation point. The maxillae are a pair of appendages with internal knife- or hook-like lobes and external segmented feelers called palpi (figs. 2.11k, 2.12k, 2.13k, 2.14k). The labium consists of paired appendages that form the lower lip and are fused together or separated toward the tip, closing the mouth from behind or beneath. A pair of segmented feelers, the labial palpi, is usually articulated near the base of the labium (figs. 2.13-l, 2.14-l).

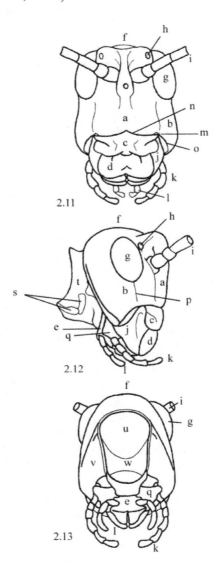

2.11 Frontal view of a grasshopper head.
2.12 Lateral view of a grasshopper head.
2.13 Posterior view of a grasshopper head.
Legend for various insect heads.

a-frons	l-labial palpi	w-tentorium
b-gena	m-anterior tentorial pits	x-muscles
c-clypeus	n-epistomal suture	y-cibarium
d-labrum	o-subgenial suture	z-rostrum
e-labium	p-subocular suture	*a*-labella
f-vertex	q-maxilla	*b*-galea
g-compound eye	r-proboscis	*c*-paraglossa
h-ocellus	s-cervical sclerites	*d*-glossa (tongue)
i-antenna. ia-scape, ib-pedicel, ic-flagellum	t-cervix	hp-hypopharynx
j-mandible	u- foramen	
k-maxillary palpi	v-post-gena	

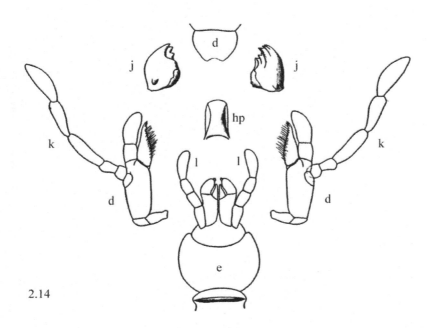

2.14

Fig. 2.14 Generalized mouthparts of a chewing insect
Refer to Figs 2.11-13 for legend.

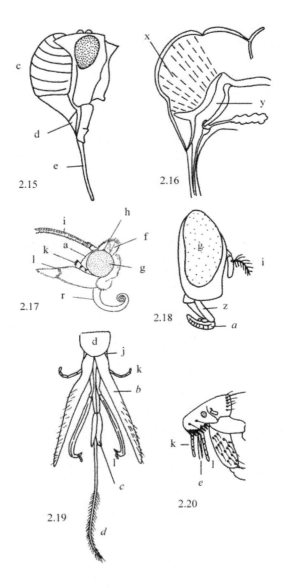

2.15 Lateral view of a cicada head.
2.16 Cross-section of a cicada head.
2.17 Lateral view of a moth head.
2.18 Lateral view of a fly head.
2.19 Mouthparts of a bee.
2.20 Lateral view of a flea head.
Refer to figures 2.11 – 2.13 for legend.

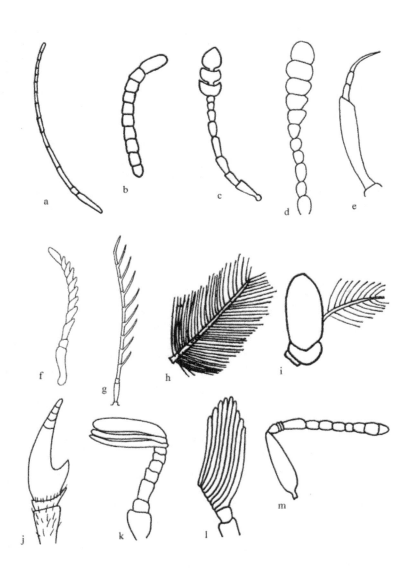

2.21 Types of antenna.

a-filiform
b-moniliform
c-capitate
d-clavate
e-setaceous
f-serrate
g-pectinate
h-plumose
i-aristate
j-stylate
k-lamellate
l-flabellate
m-geniculate

These typical insectan mouthparts are vari... lified in different orders. In some, the maxillae and mandibles re... nple chewing or grasping function, while in others a proboscis or su... e is formed from the conjoined mouthparts. Cicadas have sucking parts capable of piercing plant tissues (fig. 2.15). The clypeus is enla... ntaining muscles that contract to create a negative pressure that suc... nt fluid into the mouth (fig. 2.16). This enlarged region of the clyp... led the cibarium. Butterflies and moths have the mouthparts arrang... coiled proboscis (fig. 2.17). House flies have sponging mouthparts...), and bees have the mouthparts modified into lapping structures (2... as have modified the basic pattern into a piercing structure capable o... nto the tissues of their hosts (fig. 2.20).

The antennae (fig. 2.21), like the mouthp... exhibit a wide range of variation from simple filaments to compl... structures. The segment closest to the head is called the scape, ... followed by the pedicel. The remaining antennal segments comprise... llum.

The Thorax

The insect thorax is composed of three se... he two posterior segments are often closely united to form a pteroth... flight box." The anterior segment, or prothorax, consists of a dorsa... eferred to as the pronotum, a ventral sternite called the prosternum... uced pleurites at each side (fig. 2.10). The two posterior segments (t... and metathorax) have similar dorsal and ventral sclerites (the mes... etanota and the meso- and metasterna) and usually have well-devel... rites at the sides. The fused pleurites are usually separated into an... pisternum and a posterior epimeron by the pleural suture, which is... ted internally by the pleural ridge. The external groove and internal... he pleural suture remain in many insects that have secondarily lost th...

The Legs

Nearly all adult insects have legs or vestig... articulating with the three thoracic segments. The insect leg is usu... posed of a coxa, trochanter, femur, tibia, and tarsus (fig. 2.22). Th... s divided into as many as five secondary segments and is usually ... d by two claws, which are actually moveable spurs. This last section... arsus is called the pretarsus. In the Archaeognatha, however, thre... are present, the middle one being the smallest. This middle claw i... chment point of the main leg ligament, and is replaced in most adva... cts by a plate-like sclerite (fig. 3.2).

The legs of insects are modified for many f... such as walking (cursorial), jumping (saltatorial), grasping (raptorial)... ng (natatorial), or digging (fossorial). These modifications cut across ordinal lines, and similar types of legs may be found in several different orders.

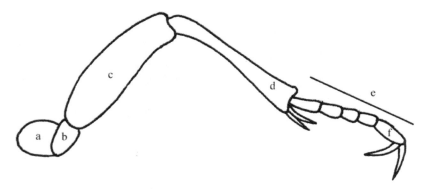

2.22 The parts of an insect's leg
a – coxa, b – trochanter, c – femur, d – tibia, e – tarsus, f - pretarsus

The Wings

The wings of insects are essentially flat, projecting appendages developed on one or more of the thoracic segments. In nymphs or naiads or during a pupal stage, they develop on the thoracic segments as sac-like flaps. Tracheal tubes and tubular formations of epidermal cells extend into these flaps, and following the last molt (the definitive adult molt), blood is pumped into the flaps, expanding them like flattened balloons. As the wings expand, the trachea and epidermal tubes become sclerotized and form the veins that support the hardened wings.

As the blood is withdrawn, the upper and lower layers of cuticle collapse upon each other so that, in a typical wing, thickened supporting veins are separated by thin membranous areas. Some blood (or hemolymph) circulation and some reparative epidermal cells remain in the wings, even when they are dried and apparently completely hardened.

Flat, membranous wings similar to those found in fossil insects of the Upper Carboniferous are a characteristic of many living insects. In others, however, the wings have been greatly modified by various processes of sclerotization. These modifications are often specific to insects of certain orders and are instrumental in their classification.

Membranous wings are clear and thin, and the veins are clearly visible (fig. 2.23), while tegmina are characterized by forewings with membranes that are parchment-like and usually translucent, with detectable veins. In insects with tegminous forewings, the hind wings are usually membranous and are protected by the forewings, which are folded over them when not in use (fig. 2.24). Elytra (fig. 2.25) serve a similar protective purpose, but are more heavily sclerotized and fit more closely to the contours of the body. They are opaque and the veins are detectable only with difficulty or not at all. Hemelytra are forewings which are heavily sclerotized at the base but suddenly membranous toward the tip, with distinct veins. Unlike elytra, hemelytra usually fold over each other as well as over the hind wings in repose (fig. 2.26).

In the Diptera, the hind wings are modified into knob-like balancing organs called halteres (fig. 2.27). Pseudohalteres (fig. 2.28) are similar knobbed structures found only in some Strepsiptera. They differ from true halteres in that they are modified from the forewings rather than the hind wings.

Fringed wings are very small, strap-like appendages with fringes of long, hair-like setae, which function as wings in Thysanoptera (thrips), some Coleoptera (beetles), Hymenoptera (bees, wasps, and ants), and Lepidoptera (moths). They are found only on very small insects (fig. 2.29).

Modifications of the wing surface are also important in classification. The wing cuticle often has many setae on its surface, which form sensory organs or are modified into protective scales or other structures. In Lepidoptera, for example, the wings may appear opaque because of the myriads of tiny, flat, overlapping scales, whereas the Trichoptera have wings that are covered with fine hairs.

Some members of several orders have secondarily lost the wings, and winglessness occurs in members of nearly all orders except the Odonata (dragonflies), Ephemeroptera (mayflies), Neuroptera (lacewings), Raphidioptera (snakeflies), and Megaloptera (dobsonflies). Winglessness is often associated with some specialized habitat or feeding habit. For example, winglessness occurs in the Grylloblattodea, which live near glaciers, and in the Psocodea (sucking and biting lice) and Siphonaptera (fleas), which are all ectoparasites of birds and mammals. The Embiidina (webspinners) have winged males but wingless females, as do some members of other orders. Some insects have winged females and wingless males, but the reverse is more common. Ants (Hymenoptera: Formicidae), Isoptera (termites), and Zoraptera have wings that are shed after a mating flight.

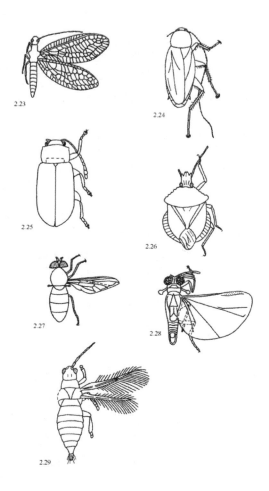

Types of wings:
2.23 Membranous wings.
2.24 Tegminous wings.
2.25 Elytra.
2.26 Hemelytra.
2.27 Halteres.
2.28 Pseudohalteres.
2.29 Fringed wings

    Ancient insects apparently held the wings straight out to the sides, slanting backward, or together upright above the back when at rest. The Odonata and Ephemeroptera retain this ancient paleopterous attitude of the wings in repose, and the condition seems to have been secondarily developed in some living butterflies and moths. The rest of the living insects fold the wings horizontally over the back when in repose by intricate manipulations of sclerites and folds in the wing. This adaptation is called neoptery. If the wings are similar in size or the hind wings are smaller, secondary folding of the latter may not be necessary. The termites usually have very similar wings that simply fold together over the back, but in an Australian family, the hind wings have an anal area which folds secondarily, so the apparently simple condition may also be secondary. The most complex patterns of folding of the hind wings occur among the Dermaptera (earwigs) and Coleoptera, where the hind wings not only fold more or less fan-like beneath the forewings, but may fold cross-wise as well. Fan-like folding beneath protective tegmina is common in the grasshoppers and roaches.

The Venation of the Wings

The venation and other features of wings are extensively used in classification. Figure 2.30 shows a hypothetical wing of a membranous-winged insect with the principal named veins, cross-veins, and wing sclerites indicated in the Comstock-Needham system. No insect, living or extinct, ever possessed all of these veins and features, but some combination of them occurs in nearly all insect wings.

It seems probable that all insect wings are derived from a single primitive type. The ancient insects possessed wings with many longitudinal veins and an intricate network of cross-veins called the archedictyon. This condition is retained, in modified forms, in the living Ephemeroptera and Odonata. The evolution of wings, however, has largely been toward a reduction of the wing venation by the loss or fusion of veins and cross-veins.

The principal longitudinal veins of a typical wing (counting from the anterior border downward and backward toward the base of the wing) are the costa, subcosta, radius, media, cubitus, postcubitus (the first anal vein), and a variable number of anal and jugal veins. In many insects, however, these veins have become so intricately modified that it is sometimes difficult to identify them.

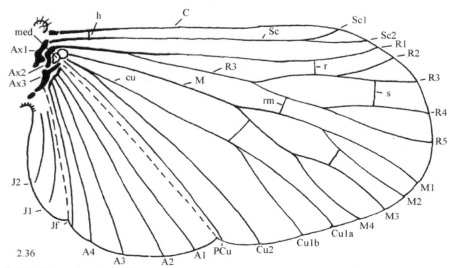

2.30. Hypothetical diagram of primitive wing venation. Explanation of lettering: Ax 1, 1st axillary sclerite; Ax 2, 2nd axillary sclerite; Ax 3, 3rd axillary sclerite; C, costal vein; cu, cubitus; cu_{Ia}, cu_{Ib}, cu_2, branches of cubitus; h, humeral cross-vein; J_1, J_2, 1st and 2nd jugal veins; jf, jugal wing fold; M, media, M_1, M_2, M_3, M_4, branches of media; m-cu, mediocubital cross-vein; m, medial cross-vein; med, medial plates; PCu, postcubital or 1st anal vein; R, radius, R_1, R_2, R_3, R_4, R_5, branches of radius, Rs, radial sector; r, radial or anterior cross-vein; rm, radiocubital cross-vein; Sc, subcostal vein; Sc_1, Sc_2, branches of subcosta; s, sectorial cross-vein; A1, A2, A3, A4, anal veins (modified after Snodgrass).

The Abdomen

The insect abdomen is made up of twelve segments, but external evidence of all segments is not present. Each abdominal segment usually has a dorsal and ventral sclerite (referred to as the tergum and sternum or tergite and sternite), but the lateral pleurites are usually reduced or hidden. The terminal segments of the abdomen are usually modified for various functions of mating, egg-laying, or defense. In many insects, a pair of lateral terminal cerci are present, and in some, a median filament is developed from the telson. Male and female genitalia are usually developed from various terminal segments of the abdomen. An ovipositor is sometimes present toward the tip of the female's abdomen. The genital structures differ strikingly in different insects, and male genitalia are widely used in classification at the species and generic levels.

The Alimentary Canal

The alimentary canal or digestive syste... sects and other arthropods is superficially simple. In insects it... divided into the foregut, midgut, and hindgut (fig. 2.31). The foregu... ts of the mouth, esophagus, crop, and proventriculus. These are all l... h epithelium and with a lining cuticle, the intima, which must be shed... ing. The midgut is not lined with cuticle. However, at the junctu... foregut and the midgut, glandular cells produce a thin, chitinous... phic membrane, which lines the interior of the midgut and pre... protects it from damage as food passes through. Diverticula called... caecae are often found at the juncture of the foregut and the midg... stion takes place within the peritrophic membrane. The hindgut is... erior end of the midgut and begins where the Malpighian tubules... hese tubules are derived from the epithelium of the hindgut in inse... re not lined by a chitinous layer, and the remainder of the hindgut ha... na similar to that of the foregut. The hindgut may be divided in ma... into an anterior ileum or colon and a posterior rectum before the an...

The intima of the foregut and hindgut are... nsclerotized, but in many insects the proventriculus is sclerotized to f... mill" or grinding organ. The rectum may also have longitudinal scler... muscles attached to help extract water from the feces. Ridges... rectum form a characteristic pattern on the surface of the excreta,... of many insects, especially in the Lepidoptera. The intima of both... gut and hindgut must be shed upon molting, along with the sclerites... oventriculus and the rectal ridges, which are often easily seen attached... ed skins.

The Tracheal System

The respiratory system of insects consists... ings in the body (fig.2.32) that connect to tubes that run the length a... of the body (fig. 2.33). The openings are called spiracles and the... e called trachea. Within the trachea are spirally or helically arrange... ting fibers called taenidia. Tracheal tubes branch and decrease i... s they penetrate inwardly, until the tracheoles are reached. The tra... ange from 0.2 to 1.0 micron in diameter and are the site of gaseou... ge with the cells. The spiracles of the main tracheal tubes may be s... enings flush with the surface of the cuticle, or they may be elabor... lified to prevent evaporation (fig. 2.32). In most, there is at least... r sclerite around the mouth of the opening and an internal vestibul... ich the air tubes connect. A closing apparatus is present in many in... l glands secreting oily substances to prevent wetting are often associat...

In insects, the simplest tracheal system is fo... e Thysanura and Archaeognatha. In the Thysanura, there are nine p... iracles, and each pair is connected by means of a group of branching tracheal tubes, which do not connect with those of other spiracles. In the Archaeognatha, there are ten pairs of spiracles, which are connected by horizontal and longitudinal tubes to the other spiracles in order to form a tracheal system. In the winged insects, the number of spiracles is reduced and the system may be specialized to include air bladders, sounding chambers, or other mechanisms. In some small insects, the tracheal system has been altogether lost and respiration is carried on through the thin cuticle.

The Insect Circulatory System

The insect circulatory system is relatively simple when compared to that of a vertebrate or even to some of the annelid worms. Its simplicity is correlated with the elaborate tracheal system, which carries oxygen to the tissues and removes much of the carbon dioxide. The main role of the circulatory system is the distribution of the necessary carbohydrates and fats needed for energy production and of the amino acids necessary for growth. It also functions in the transport of hormones, vitamins, and other products needed for the functioning of tissues and organs and carries a variety of blood cells that perform many complex functions.

Insects have an open circulatory system with a single dorsal vessel that pushes the insect's blood towards the brain. The abdominal portion of the dorsal vessel is called the heart. It has paired openings, or ostia, along the sides, through which the blood can be drawn or discharged. The ostia may be simple slits, or may be modified into elongate tubules or elaborate valves. Circulation of the blood is accomplished by rhythmic peristaltic contractions of the muscle layers forming the heart. Usually, blood is drawn into the ostia and propelled into anterior portions of the dorsal vessel, which is called the aorta. The aorta may take on a variety of forms, but in general it simply discharges into the head. Blood flow can also be redirected by the reversal of the peristaltic movements of the heart, so that blood may be discharged into the abdomen.

The circulation through the legs, wings, and other parts of the body may be aided by pulsatile organs, which serve as accessory hearts and pump the blood from the larger sinuses of the body cavity into smaller ones. The rate of circulation to various tissues varies from time to time in correspondence with the physiological state of the insect. The hemocoel (fig. 2.34) is usually divided by membranes into three main parts: the perivisceral, pericardial, and perineural sinuses.

The blood, or hemolymph, contains a variety of blood cells known as hemocytes, which may circulate in the plasma or be attached to the walls of the dorsal vessel or sinuses. The blood cells ingest small particles, surround parasites, facilitate transport of food materials, and aid in the coagulation of the blood and sometimes in wound healing. Nephrocytes are closely related

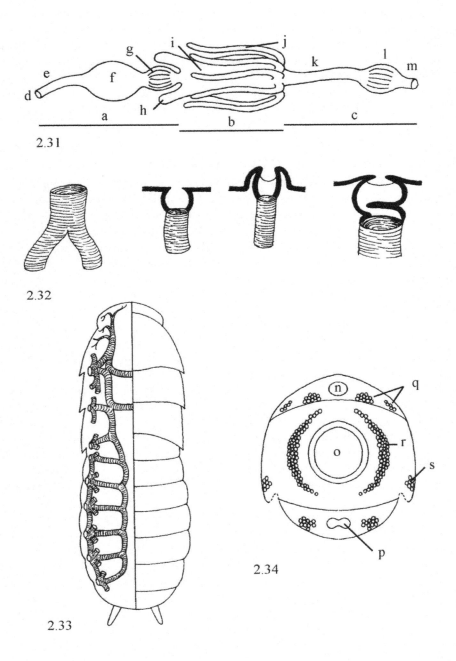

2.31 Stylized insect digestive system.
2.32 Modification of spiracles from simple to complex, showing the vestibule.
2.33 Tracheal system of a cockroach.
2.34 Cross-section of an insect showing features of the circulatory system, with the pericardial sinus (top section), perviceral sinus (central section), and the perineural sinus (bottom section).

a-foregut	g-proventriculus	m-anus
b-midgut	h-gastric caecae	n-dorsal aorta
c-hindgut	i-ventriculus	o-gut
d-mouth	j-Malpighian tubules	p-ventral nerve cord
e-esophagus	k-ileum	q-nephrocytes
f-crop	l-rectum	r-fat body cells
		s-oenocytes

to blood cells. They function in the breakdown and redistribution of various substances found in the blood. Oenocytes are another type of cell found in the blood or in the fat bodies; their function is not well understood, but they may play a role in cuticle production.

The Excretory System

The principal excretory organs are Malpighian tubules and the rectum (fig. 2.31). Malpighian tubules perform the major excretory functions of removing nitrogenous wastes from the hemolymph in most insects. They also function in maintaining water and salt balance. The tubule system may simply be a set of blind tubes formed by living cells, or the ends of the tubes may be embedded in the wall of the rectum, where they presumably reabsorb material. In some insects, solids as well as materials in solution may be taken up by Malpighian tubules. The wastes and water then move to the rectum, where water is reabsorbed. The nitrogenous wastes can then be expelled with the feces.

The fat bodies may also play a role in storing wastes. They usually form just beneath the cuticle and around the digestive tract and other internal organs. Fat bodies store food materials, such as fat, carbohydrates, and proteins, but they also aid in removing urates or calcium salts from the blood in various insects. These wastes are stored and may later be excreted by the Malpighian tubules. The fat bodies increase or decrease in size with the state of nutrition of the insect, but even when depleted, their cells remain as an irregular lining of the body cavity.

CHAPTER 3: MUSCLES, LOCOMOTION, AND FLIGHT

Insect muscle is similar in some ways to vertebrate muscle, but differs in certain basic attributes. Unlike vertebrate muscle, insect muscle is usually colorless or gray because most insects do not have myoglobin or other respiratory pigments in the muscles. Insects possess only striated muscle (produced by thick myosin and thin actin filaments that give the muscles a banded appearance), whereas vertebrates have smooth muscle in addition to striated muscle.

The action of insect muscles is the same as in vertebrates in that the muscles always pull and never push. Thus, muscles are arranged in an antagonistic system: one muscle pulls a body part in a certain direction, and another muscle comes into play to pull it in the opposite direction. This pulling is the result of the contraction of muscle fibers or muscle cells. However, because insect muscles are inside the exoskeleton, many movements involve complex arrangements of muscle attachments. The muscles of insects are not stronger than those of vertebrates, even though insects are capable of amazing feats: fleas can long-jump nearly 200 body lengths, and some insects beat their wings 1,000 times per second. These feats are possible because insects are small, and the amount of power generated compared to the mass of the animal increases as the animal's body decreases in size.

The arrangement of insect muscles is complex—so much so that in the eighteenth century, Pierre Lyonnet, a French biologist, created a controversy by demonstrating that there were more muscles in the larva of the goat moth than in the human body. The arrangement differs in various groups of insects, so we will discuss only a few muscles, the actions of which can be readily demonstrated.

Locomotion

Squirming or wriggling locomotion of caterpillars results from progressive waves of peristaltic movement in trunk muscles. Looping is a modification of wriggling in which the front and rear ends of the animal are alternately fixed by grasping the substrate as the other end moves forward. This means of locomotion is best exemplified by inchworms, which are the larvae of geometrid moths.

Cursorial insects with hardened exoskeletons walk by moving three legs (the left front and hind legs and the right middle leg) simultaneously as a tripod with the tarsal claws anchoring the leg to the surface. Then the remaining legs (the left middle and right front and hind legs) move together to complete the stride. These alternating tripods result in a very stable stride because the insect's center of gravity is within these moving tripods. Grasping and digging result from modifications of walking movements, whereas swimming is accomplished by different rowing movements in different groups of insects.

Functioning of the Leg Muscles

The diagram of the jumping leg of a grasshopper (fig. 3.1) illustrates how the leg muscles function. The muscles in a walking leg are similar, but the roles of the levator and depressor muscles of the tibia are reversed and the muscles are less different in size. (Generally, levators move a structure upward and depressors move it downward.) In addition to the muscles shown, there is a complex of muscles that move the coxa.

Some insects have the trochanter separated from the femur, and there is a muscle that retracts the femur upon the trochanter. In the grasshopper, however, there are no femoral muscles, and the femur is moved by the trochanter, which is fused to the base of the femur. The muscles of the trochanter originate in the metathorax or on the inner surface of the coxa and attach directly on the trochanter. Muscles 1 and 2 are depressors which, when contracted, rotate the trochanter and femur downward on the joint of the trochanter with the coxa. Muscle 3 is a levator that rotates the trochanter and femur upward.

The muscles that move the tibia are in the femur, and since jumping requires a great deal of power, the levator is large. The levator of the tibia (4), which raises the tibia, also causes it to move backward. When the tibiae of both sides are being affected by these large muscles with the tibia and claws planted on the ground, the whole insect moves up and forward—that is, it jumps. The levator muscle (4) inserts on a large apodeme (tendon) that extends into the femur from the upper part of the tibia, which swings on a double lateral joint on the femur. A long apodeme extends from the

pretarsus through the tibia into the femur, and mu   rs from the large
femoral levator (4) also insert on it, aiding in fu   ing of the tibia.
Muscle 5 is the depressor of the tibia, which inse   smaller apodeme
extending inward from the lower part of the tibia.   he tibia, 6 and 7
are the depressors of the pretarsus, which cause   to dig into the
surface on which they are sitting. The long tendo   not insert on the
claws themselves, but on a small plate beyond them   s a vestige of the
third claw found in more primitive insects such as   inura. Muscles 8
and 9 are the levators and depressors of the tarsu   which does not
have individual muscles in its segments. In jumpin   quence of muscle
contractions is approximately as follows: 1-2, 4, 6-7 n the leg refolds,
the sequence is 8, 5, 3.

    In the walking legs, muscle 5 is larger than   , since walking is
produced by lifting the leg, rotating it forward, and   ging it down and
pulling the body forward. The sequence in wa ll   thus: the leg is
extended to a new position, the tarsus is pressed d   btain a hold, the
leg is depressed to pull the body forward, and the tar ted to release the
hold of the claws.

    In walking, the hind legs of a grasshopper   minor assistance,
and the main weight is borne by the anterior pairs.   mber the legs on
each side: R(ight) 1,2,3 and L(eft) 1,2,3, then the s   n jumping is R3-
L3, but in walking it is R1-L2, R2-L1, etc. As w   lready discussed,
insects that walk on all six legs fall alternately from od of legs to the
other: R1-L2-R3 to L1-R2-L3 and so on.

    The musculature of the coxal base diff   fferent types of
articulation, but in all cases opposing (antagonistic) muscles originate
within the thorax and insert within the coxa to pr   ariety of actions.
The lower end of the pleural suture serves as a fu   r the coxa. The
single-fulcrum type of coxal articulation has th   est flexibility of
movement, but the other types, with double or t   iculations, allow
greater power to be applied in a specific direction   grasshopper, for
example, the muscles that move the hind femur ar   in the coxa and
thorax and insert in the trochanter, which is fused   base of the femur
(see fig. 3.1). The muscles that raise and lower th   nter also raise or
lower the femur. The large muscles that extend th   bia and tarsus are
located in the femur and insert on tendons that ext   o the sclerite just
behind the planta (fig. 3.2). Small muscles within th   so help to extend
or retract the tibia and the jointed tarsus.

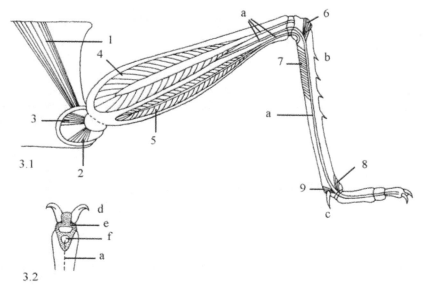

3.1 Anatomy of a grasshopper hindleg.
3.2 Pretarsus of a grasshopper.

1-depressor of trochanter	5-depressor of tibia	9-depressor of tarsus	d-claw
2-depressor of trochanter	6-depressor of pretarsus	a-tendons	e-planta
3-levator of trochanter	7-depressor of pretarsus	b-spines	f-plate to which retractor tendons attach
4-levator of tibia	8-levator of tarsus	c-spur	

The Origin of Wings

Several hypotheses have been proposed for the origin of wings. One is that wings developed from paranotal lobes or flaps, which extend backwards from the thoracic nota. This is clearly the way in which wings develop in the nymphs of winged insects, and the hinging sclerites and muscles can be shown to have been derived from portions of the notum and from the primitive thoracic and leg muscles. The paranotal lobes may have evolved the articulations and muscled before they were ever used as wings. They may, for example, have functioned as protective coverings for soft-bodied insects, or they may have been used as supplementary appendages in climbing. Evidence for the latter exists in living butterflies, which, upon emergence from the pupa, may use the wing rudiments before the wing is completely unfolded. The primitive venation of the wing may at first have been primarily useful for support in climbing and for protection of the wing against abrasion. Later, the wing venation may have assumed additional protective value in simulating the venation of plant leaves to confuse predators, as suggested by the amazing resemblance between the venation of some ancient insect wings and the venation of fern leaflets. It has also been suggested that the original paranotal lobes may also have functioned as sexual attractants and holdfasts in the transition from external spermatophore transfer to copulation. It is also possible that paranotal flaps could have functioned in thermoregulation.

Another hypothesis is that wings developed from gills, such as those found in the nymphs of many living aquatic insects, but that notion is faced with a number of difficulties. Wings do not occur in the immature stages of any living forms, but the gill theory implies that it was the nymphal stage which developed wings and, hence, flight. Also, the gills of such an aquatic nymphal stage would require a secondary return of the terrestrial Thysanura-like ancestor to an aquatic habit. Some living Thysanura, however, do have flattened lobes on the thoracic segments, which are apparently an adaptation to creeping in closely confined spaces. These lobes strongly support the theory that wings arose from similar paranotal flaps.

A third hypothesis is that wings evolved from parts of legs. This hypothesis came from the discovery that certain receptors are only found on legs and wings and not on the surface of the body. Moreover, the genes that are used in embryonic insects to produce legs are also used in the production of wings. Indeed, the genes that produce the lateral outgrowths on primitive wingless insects are also used in wing production, suggesting that insect wings may have evolved from the interaction of the genes that would produce paranotal lobes and genes that would produce legs.

At the start of the Upper Carboniferous deposits, winged insects appear in considerable numbers. These, however, are already specialized winged insects, so we have no fossil record of the transition from a wingless condition to true flying forms. We can surmise, however, that flight began first as simple gliding when the primitive insects fell or were chased from their position on plants. This aerial advantage would have had tremendous survival value in avoiding the attacks of predators. The development of basal sclerites and muscles to flap the wings added another dimension to wing function. Once flapping flight was achieved, the insects underwent a tremendous burst of evolution. Before the end of the Upper Carboniferous, several of the modern orders were already established, and most of the living orders were present or foreshadowed before the end of the Permian.

The evolution of wings probably occurred in the late Devonian and early Carboniferous. By the Upper Carboniferous, the winged insects were already clearly divisible into two major groups: the paleopterous insects, in which the wings extended straight out to the sides and were incapable of folding flat over the back, and the neopterous insects, in which the wings, by means of secondary hinges, could be laid overlapping on the back when not in use in flight.

Flight Muscles

The flight muscles are usually divided into two classes: direct muscles (figs. 3.3, dm), which attach to sclerites in the membrane of the wing base (fig. 3.4), and indirect muscles (figs. 3.3 and 3.5, lm and tsm), which attach within the thorax and exert their force on the thoracic walls rather than directly upon the wings. Both classes of flight muscles function in moving the wings. The direct muscles were probably the main muscles involved in flight in the ancient insects, because the Odonata, an ancient order, still use direct muscles to generate the power stroke of the wings. However, even these insects produce the upstroke of the wings with indirect muscles. The higher insects use the direct muscles only to orient the wing while the indirect muscles produce the power stroke in addition to the upstroke.

In the higher orders of insects, the power stroke in flight is produced by enlarged longitudinal muscles that run from the anterior end of the mesothorax to the anterior end of the segment behind the metathorax, and by lateral oblique muscles that originate on the posterior ends of a notum and attach to the inner dorsal surface (figs. 3.6). In the higher insects, the meso- and metathorax are largely fused together to form the pterothorax, or flight box. When the longitudinal muscles contract, the nota are arched upward and the wings are pressed downward because of the structure of the sclerites and membrane at their bases (figs. 3.6, 3.7). The upper end of the pleural suture is the fulcrum on which the wing swings.

The moving of the wings upward is produced by the contraction of the tergo-sternal muscles that oppose the longitudinal muscles (figs. 3.8, 3.9). Contraction of the tergo-sternal muscles depresses the nota, which press down upon the wing bases, causing the wings to swing up on the pleural fulcra.

Although insect wings are essentially flat, the upper surface is slightly convex and acts as an airfoil in flight. As in the airfoil of a jet, air must travel farther over the top of the wing than the underside, resulting in the production of lift. Unlike a jet's fixed wings, however, insect wings beat in a figure-eight pattern. The front edge is pulled down and forward by muscles attaching to sclerites in the anterior part of the lower membrane near the wing base (the basalares) (fig. 3.4). Immediately thereafter, the back edge is pulled back and down by muscles attaching to sclerites in the posterior part near the lower wing membrane (the subalars)(fig. 3.4). The beating of the wing actually increases the curvature of the wing, producing additional lift, and the figure-eight pattern of the wing movement pushes against the air, resulting in thrust (forward movement). This changing wing curvature, combined with the movements of the wing, produces a kind of rowing action that creates a helical airflow along the leading edge of the wing and forms vortices, or cylinders of air, that are shed from the wing tip, resulting in additional lift.

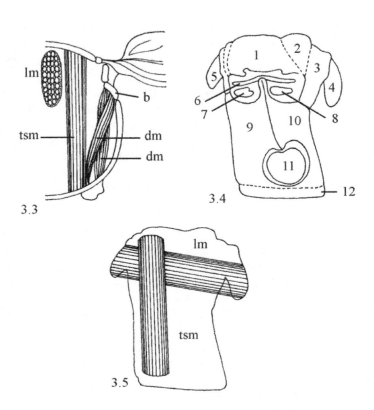

3.3 Internal view of direct and indirect flight muscles.
lm-longitudinal muscle b-basalare
tsm- tergo-sternal muscle dm-direct muscle

3.4 Framework of a wing-bearing thoracic segment.
1-scutum of the notum   5-prephragma   9-episternum
2-scutellum   6-wing   10-epimeron
3-postnotum   7-basalare   11-coxa
4-postphragma   8-subalare   12-sternum

3.5 Indirect flight muscles
lm-longitudinal muscle   tsm-tergosternal muscle

The biomechanics of all the refinements of insect flight are not yet fully understood. Some insects can hover, soar, dart, turn somersaults in the air, change directions with incredible swiftness, and perform many other remarkable aerobatics. These feats are largely controlled by the action of direct muscles and coordinated by the nervous system.

A considerable amount of the power for flight is stored in elastic proteins, or resilins, in the membranes of the wing bases. These proteins are unique in being almost perfectly elastic. When the wing is raised, they are deformed in such a way that on the downstroke, they contribute additional force on the wing. Another form of energy feedback is the click mechanism, which, in some insects, prevents the down-sweep of the wing until considerable force is generated. The click mechanism then releases the force all at once and the down-sweep is shortened in duration, but its force is increased.

One problem in insect flight is the turbulence produced by the forewings interfering with the hind wings and decreasing the efficiency of flight. This problem is compensated for or avoided by a number of interesting devices. In the Diptera, only the forewings function as lifting surfaces, and the hind wings are reduced to tiny balancing organs called halteres. In the Coleoptera, the hardened forewings (elytra) are held upward and the hind wings furnish the power for flight. In wasps and some moths, the wings are joined by interlocking hooks, which function almost like a zipper that holds the two wings together, forming a single lifting surface. In other insects, strong setae at the wing bases hold the wings together. In the butterflies, the wings overlap broadly and thus function as a single airfoil. The Odonata avoid the forewing turbulence by keeping the two wings out of phase with each other, allowing the hind wing to engage the vortex created by the forewings to produce additional lift during hovering.

When winged insects are not flying, they hold their wings in two different ways. The earliest flying insects simply held them extended to the sides or raised them together over the back (the paleopterous positions). Most modern insects, however, are able to fold the wings flat or roof-like over the back (in neopterous positons) when they come to rest. This wing folding permitted insects to crawl into small spaces without damaging the delicate wings, which opened up many more ecological opportunities for the little fliers.

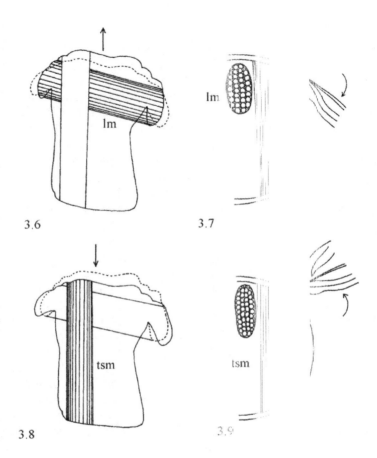

3.6 Lateral section of an insect with the longitu[dinal mu]scle contracted, which results in pushing the notum [up.] Dashed lines show the position of the notum at rest.

3.7 Cross-section of the longitudinal muscle c[ontracted], causing the notum to be pushed upward, resulting in the do[wnward] movement of the wing.

3.8 Lateral section of an insect with the tergo-s[ternal m]uscle contracted, showing the pulling down of the no[tum. Da]shed lines show the position of the notum at rest.

3.9 Cross-section of the insect with the tergo-st[ernal mu]scle contracted, causing the notum to be pulled dow[n, w]hich in turn pushes the wing upward.

CHAPTER 4: THE NERVOUS SYSTEM

The central nervous system in insects shows an elaboration of the basic pattern found in the all arthropods and annelids, which may be related to the common arthropod ancestor. In insects, the central nervous system includes a brain, which is located in the head, and a ventral nerve cord that incorporates enlarged masses called ganglia. When combined with the sense organs, it provides the insect with the ability to sense the world around it and respond appropriately.

The changes in the nervous system from the primitive wingless insects (fig. 4.1) to the higher winged insects (fig. 4.2) largely involve increases in complexity of the central nervous system and the fusion of the ganglia of the ventral cord. The extreme is reached in such insects as the house fly (fig. 4.3), where only the brain, subesophageal ganglion (a ganglion in the head below the esophagus), and a single ganglionic mass, formed from the fused thoracic and abdominal ganglia, occur.

The general functioning of the nervous system is similar in all animals. It is based on the transmission of nerve impulses through specialized cells called neurons. Neurons (fig. 4.4) are made up of cell bodies with branches that receive stimuli (dendrites), and branches that send impulses away from the cell body (axons). Neurons do not come into contact with each other; there is a gap, the synapse, between the axon of one neuron and the dendrite of the next neuron. A neurotransmitting molecule moves across the synapse and in turn triggers the impulse of the receiving neuron. Insects have several neurotransmitting molecules, with acetylcholine being the most common.

There are three types of neurons, based on their function. The sensory neurons receive information from the insect's surroundings and send it to the central nervous system. Interneurons interact with other neurons, relaying impulses from one neuron and sending on to the next. Motor neurons transmit impulses from interneurons and in turn trigger a muscle to contract. They are critical in the insect's response to stimuli.

Certain types of basic nerve impulses are generated within the ganglia of the central nervous system and accomplish routine activities such as breathing, contraction of the heart, and tissue metabolism. Ordinarily, however, an insect remains inactive unless it receives stimuli from external sources.

Function of Nerves

A nerve impulse is the electrochemical message of the nerves, and is basically the same in all animals, from flatworms to humans and the insects. The axons carry nerve impulses as an "all-or-none" process. That is, an axon is either carrying a nerve impulse, or it is not carrying a nerve impulse. Thus, the effect a neuron has on the tissue to which it transmits the electrochemical message can only be varied by changing the frequency of nerve impulse conduction. An axon may "fire" only a few times per second or up to about 1,000 times per second. The neurons associated with sense organs may also increase the frequency of the impulses transmitted to the nervous system depending upon the level of stimulation. For example, a "blinding light" may excite the eye to such a degree that the frequency of transmission becomes so rapid that an image can no longer be formed.

A resting neuron contains more potassium ions inside the cell and axon than outside, where sodium and chloride ions predominate. In the fluids around the neuron, sodium may be ten times as common as inside the cell; conversely, potassium may be 25 or more times as common inside the cell than outside. This creates a potential difference, or a voltage between the positive outside and negative inside. This is called the resting potential, equal to about -70 millivolts.

A nerve impulse is a rapidly moving change in this electrochemical potential. That is, the polarization of the cell membrane changes so that the cell usually becomes positively charged inside and negatively charged outside. This is the action potential. What actually seems to happen is that the cell membrane changes its permeability so that the cell becomes more permeable to sodium than to potassium, and in rushes sodium. Once started, this process is self-propagating and a nerve impulse moves along the neuron. After the passing of an impulse, the membrane quickly restores itself to the former condition, ready to react again and now practically impervious to sodium.

Only a very small amount of sodium actually enters the cell, but this must be removed. This is accomplished by the "sodium pump," which involves proteins in the membrane of the neuron. These proteins, with the aid of adenosine triphosphate (ATP) as an energy source, remove the sodium ions from inside the cell and dump them outside. As the sodium moves out, potassium is brought back into the neuron.

When the impulse reaches the end of the axon, vesicles release the neurotransmitting molecule, acetylcholine in many cases, into the synapse. The neurotransmitting molecules couple with the membrane in the ends of the dendrites. This, in turn, triggers the next neuron to "fire" and the process begins again. An enzyme, such as acetylcholinesterase, may remove the neurotransmitting molecules from the dendrite, which shuts off the firing of the next nerve.

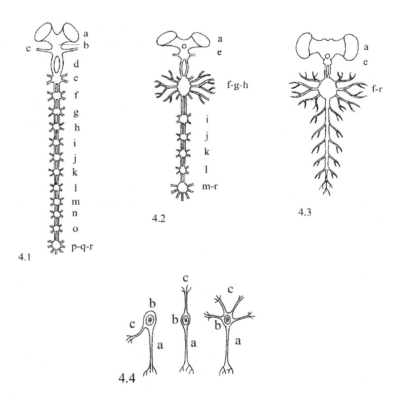

4.1 The central nervous system of *Machilis* (Archaeognatha).
4.2 The central nervous system of *Stratiomya* (Diptera).
4.3 The central nervous system of *Musca* (Diptera).

a-optic lobes
b-brain
c-antennal nerve
d-paraoesophageal commisure
e-suboesophageal ganglion
f-g-h- thoracic ganglia
i-r abdominal ganglia

4.4 Examples of insect neurons.
a- axon, b- cell body, c- dendrite

45

The simplest conceivable nervous reaction is a reflex arc. A reflex arc occurs when a number of neurons, including a sensory neuron, transmit a nerve impulse from some sense organ into a central ganglion. Within the ganglion, interneurons transmit the impulse to a motor neuron. The motor neuron transmits an impulse to a muscle that is stimulated by electrical and chemical forces and then contracts, producing some "purposive" action. Even a very simple reflex, however, involves many neurons. All reflex actions are "purposive" in the sense that they have some survival value to the organism. Basic reflex actions can be transferred within the central nervous system. For example, if a hind leg is removed from an aquatic beetle (Dytiscidae), swimming will be adjusted by using the middle leg. Similarly, a cockroach will hold and clean its antennae with a middle leg if the foreleg is removed.

Reflex actions may also be combined in chains. A dragonfly resting on a bush can be stimulated to fly by pinching the tip of the abdomen with forceps. This is the sequence of actions: 1) the abdomen is bent downward, 2) the legs release their hold on the substrate, and 3) the wings begin to beat up and down.

The behavior of insects is largely a matter of reflex actions that are stimulated and inhibited. The brain is an important center of inhibition and allows for varying responses. If not for the inhibitory actions, the insect would be totally incapable of functioning because it would react without any sequence or "purpose" to every stimulus received. The importance of inhibition is shown by removal of the brain. If the brain is removed, dragonflies can no longer fly because they cannot release the hold of the legs.

Simple walking or flying also requires complex alternations of stimulation and inhibition, but because they can be performed even by decapitated insects, these reactions are clearly coordinated within the thoracic ganglia. Since most walking insects are hexapods, they walk by alternating a tripod composed of two legs on one side and one of the other with the opposing tripod, but if a leg is removed, a new sequence of walking must be adopted. Removal of still other legs imposes new patterns, and so it is impossible to assume that all insect activity is the result of completely preformed reflexes. The factors are complex, involving the central nervous system, muscle tonus, and other variables.

The suboesophageal ganglion is an important motor center for control of the mouthparts and also affects motor coordination of the entire insect, because normal walking does not occur after its removal. The suboesophageal ganglion also exhibits inhibitory activities. For example, a male mantid copulates more freely after removal of the suboesophageal ganglion. An insect whose brain has been removed without damage to the suboesophageal ganglion may live for months if it is fed, but it has no powers of seeking food.

The ganglia of the ventral nerve cord are centers of reflex action, combining both motor and sensory functions. Individual ganglia show considerable autonomy, controlling one side of a segment without affecting the other. Severing the nerve cord in front of or behind the prothorax allows leg movements and other actions, but there is no transfer of impulses across the break. The last abdominal ganglion, which is complex, controls the actions of the genitalia and reproductive activities.

The insect nervous system also includes nerves that lie outside the brain and the ventral nerve cord. The esophageal visceral system controls the heart and swallowing actions, and the ventral visceral system controls the action of the spiracles.

Sense Organs

Nerve cells in general are not capable of receiving direct stimuli from outside the body. Some sensory neurons may receive stimuli directly from the surface or the tissues about them, but most sensory cells are associated with specialized sense cells or organs derived from the epidermis.

Mechanoreceptors are specialized structures that respond to mechanical changes, such as pressure or vibration. They may take on a variety of forms, but all function in essentially the same manner. They are concentrated on antennae and other parts of the body, producing especially sensitive areas. The simplest mechanoreceptor sensilla are tactile hairs (fig. 4.5), which are setae provided with a sensory neuron that carries sensations into the central nervous system as nerve impulses. Another form of mechanoreceptor, the campaniform sensilla (fig. 4.6), consists of a single receptor cell that is connected to a cuticular cap. Campaniform sensilla detect stress on the cuticle. Not surprisingly, they are found near joints in the legs, mandible bases, and wing bases, where they can detect strain on the cuticle and aid in preventing injury.

Chemoreceptors, which respond to chemical stimuli, are also developed from setae. Like the tactile hairs, they are concentrated in certain parts of the body. For example, the bristles of a fly's tarsi are especially sensitive to stimuli involved with food as well as movement, and they can differentiate not only various foods, but also the concentration of sugar and other substances. Chemoreception includes two senses: gustation, which is the detection of a solid or a liquid, is synonymous with tasting, and olfaction, the detection of a gas, is the sense of smell. There are two types of hairs associated with chemoreception that differ in their morphology. One type (fig. 4.7) has many pores along the shaft of the hair or peg and is mainly involved in olfaction. The other chemoreceptor group has a single pore on a hair, peg, plate, or depression in the cuticle (figs. 4.8, 4.9, 4.10). A common feature of all chemoreceptors is the presence of sory cells within each structure.

Sound receptors involve specialized orga l scolopophores. These may be grouped together to form chordoto ns (fig. 4.11) and may be associated with tympani, which are region ible cuticle that resemble eardrums. When sound waves hit the t they cause it to vibrate, which is detected by the receptor cells d to it. In the grasshopper, for example, the tympani are loca he sides of the abdomen just behind the thorax. In the katydids, the tympani are located on the forelegs.

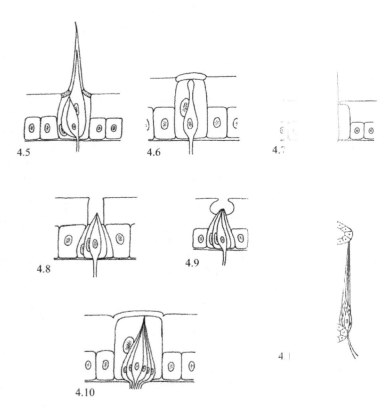

4.5 Trichoid sensilla for mechanoreception.
4.6 Campaniform sensilla.
4.7 Trichoid sensilla for chemoreception.
4.8 Ampullaceous sensilla.
4.9 Coeloconic sensilla.
4.10 Placoid sensilla.
4.11 Chordotonal sensilla.

Most of the receptors discussed are trichoid hairs or modifications of the hair. To make matters more complicated, there is no correlation between the type of hair and the function. Some structures, such as the campaniform sensilla, are always associated with stress reception, but other receptors can have more than one function. The blow flies, for example, have single hairs that contain four chemoreceptors and one mechanoreceptor, therefore serving multiple functions.

Organs of Vision

Both simple eyes and compound eyes are found in many insects. The optical mechanisms of both types develop from the cuticle itself and from the underlying epithelial cells. The sensory portion of the eyes is formed by special sensory cells.

The simple eyes of insects are of two general types: the stemmata, which are found in the larval stages of many insects, and ocelli, which occur in adults and the nymphs or naiads of insects, but not in larvae. The stemmata (fig. 4.12) usually have convex lenses formed by epidermal cells called cornagen cells. The lens, which is composed of cuticle, is hardened and transparent. The cornagen cells differ from ordinary epithelial cells primarily in being extremely translucent. Directly beneath the cornagen cells are a number of retinule cells that form a photoreceptor called the rhabdom. The stemmata are found on the head capsule of holometabolus larvae such as caterpillars, and they only see a limited portion of the larva's world. The larva apparently can accommodate this limited worldview by moving its head to obtain a more complete picture.

The ocellus (fig. 4.13) differs from the stemmata mainly in having many more retinule cells beneath a single lens. They are found in sets of three or two on the top of some insects' heads. The ocelli do not produce an image, but may function during flight to detect the horizon, allowing the insect to control its pitch and roll.

The compound eye is one of the most remarkable features of the insects. It may have originated from the massing together of ocelli, but there seem to be no survivors showing intermediate stages in its evolution. When viewed from the outside, the insect compound eye appears as a closely packed group of tiny convex lenses. Each lens corresponds to an underlying structure called an ommatidium (fig. 4.14). When seen from the side in a transverse section, the lenses appear lozenge-shaped and convex on both faces. Beneath them are the two specialized epidermal cells (cornagen cells) that secrete the lens; as in the ocelli, the cornagen cells are very translucent. Four cells below the cornagen cells combine to form the crystalline cone. These cells are also extremely translucent and their nuclei are indistinct. The crystalline cone focuses light on the retinule cells below it, which combine to form an elongate light-sensitive rhabdom made up of microvilli found on the retinule cell membranes. Six pigment cells form a covering around the crystalline cone, and accessory pigment cells overlap the retinule, shutting out light from the neighboring ommatidia. Each ommatidium is connected by axons to an optic nerve that leads to the large optic lobes of the brain.

The structure of the optical part of the eye is such that an inverted image may be formed on the rhabdom of each ommatidium, or only a difference in light intensity may be registered. Only a part of the object being viewed is focused in a single ommatidium, and the entire image is probably made up of a mosaic that is not inverted. This is known as the theory of mosaic vision. If the lenses of a compound eye are removed and used as the lens of a camera, an image is obtained, but it is made up of many minute spots, somewhat like the printed points that make up a newspaper picture. This is thought to indicate that insects do not receive dozens or hundreds of tiny pictures, but a single image made up of many small parts.

The visual acuity of an insect, such as a migratory locust or blow fly, is three to four times what can be expected from the theory of mosaic vision. Some researchers think that this indicates that a series of diffraction images may be formed at different depths into the eye. The insect reacts to one or perhaps several of these, or perhaps no image such as we recognize is formed at all. In any event, what goes into the insect brain from the compound eye are nerve impulses indicating, by their frequency, differences in light intensity. The brain must make its own synthesis of these impulses from the frequency of the impulses and the pattern in which they are received.

The diffraction-image theory is supported by the differences in day and night adaptations of the eye. The crystalline cone and the pigment cells combine to form a lens cylinder. The path of light passing through this cylinder can be changed by the migration of pigment granules in the surrounding pigment cells. In bright light, the pigment granules spread throughout the pigment cells and form a tight curtain around the crystalline cone and the upper part of the retinule cells. This focuses all the light coming into the eye directly on the distal end of the rhabdom and a single sharp image is formed. This is known as apposition. At night or in the dark, however, the pigment concentrates near the top or bottom of the respective cells, and light coming into the ommatidium is transmitted laterally to the rhabdoms of other ommatidia, as well as directly down onto its own rhabdom. This is known as superposition. This destroys or greatly distorts image formation, but greatly increases the sensitivity of the eye to dim light and allows the insect to detect movement at reduced illuminations.

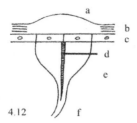
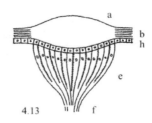
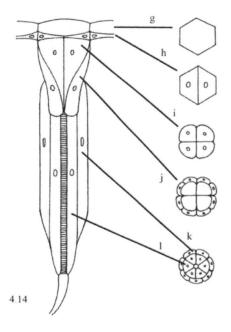

4.12 Stemmata.
4.13 Ocellus.
4.14 Ommatidium.

a-lens formed from cuticle
b-ordinary cuticle
c-epithelial cells of epidermis
d-rhabdom
e-retinule cell
f-sensory nerves
g-lens
h-cornagen cells
i-crystalline cone
j-pigment cells
k-accessory pigment cells
l-retinule cells forming the rhabdom

Research has shown that some beetles perceive motion by correlating the arrival time of changing light signals in the brain and integrating this information to judge speed and direction. A ground-speed altitude indicator that makes use of this principle was developed for aircraft. Photoelectric cells in the nose and tail of the craft indicate speed and altitude by calculating the time it takes a light signal to pass between the two cells.

Insect eyes are more reactive to the upper part of the visible spectrum and very unresponsive to the lower wavelengths; therefore, many insects perceive light in the ultraviolet range and are attracted to UV sources, or "blacklights." Conversely, ants and other insects can be observed as if they were in the dark by using red light, because their eyes do not respond to light of this wavelength. Bees and other insects can discriminate between different colors, particularly in the yellow-to-blue range. Thus, the majority of bee-pollinated flowers are blue or yellow. Red clover flowers also have a blue element. White flowers are generally pollinated by beetles, which do not discriminate between colors. When viewed with a UV camera, many flowers are revealed to have unusual patterns on their petals that help lure pollinators. We, of course, do not know what a color actually looks like to an insect, and some flowers must appear very different to them than they do to us.

CHAPTER 5: INSECT REPRODUCTIVE SYSTEMS AND DEVELOPMENT

The overall success of insects is tied to their efficiency in reproducing more insects. One of the reasons that insects are so prolific is that their unique development, with larval forms differing so greatly from adult forms, enables many immature insects to avoid competing with their parents for resources.

Primitive insects probably had a number of similar ovaries and testes, which were segmentally arranged. This is suggested by the arrangement of reproductive organs in some primitive insects, but all living forms have the ducts consolidated, and in most insects, there is only a single opening to the outside. In the mayflies, both male and female reproductive systems open through a pair of ducts and openings. The correlated parts of the male and female reproductive systems are listed in Table 5.1.

Table 5.1. The reproductive organs of the male and female insects.

Male	Female
1. Paired testes with one to many sperm tubes	1. Paired ovaries of one to many ovarioles
2. Paired ducts (vas deferens)	2. Paired ducts (oviducts)
3. Sperm sac (seminal vesicle)	3. Egg storage (calyx)
4. Median ejaculatory duct	4. Common oviduct and vagina
5. Accessory glands	5. Accessory glands
6. (No correlative structure)	6. Spermatheca and spermathecal gland
7. Genitalia (aedeagus and associated parts)	7. Ovipositor (if present)

The testes (fig. 5.1) are the site of spe... ...ction, and they connect to the seminal vesicle via the vas defer... ...e seminal vesicle serves as a sperm storage organ where material ... from accessory glands surrounds the sperm, forming the seprmato... ...is material feeds the sperm and keeps it alive until it is transferred ... ...nale through the aedeagus.

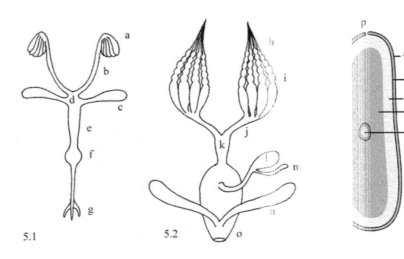

5.1 Male reproductive organs.
5.2 Female reproductive organs.
5.3 Typical insect egg.

a-testis	i-ovarioles	q-cho...	
b-vas deferens	j-calyx	r-vitel...	brane
c-accessory gland	k-common oviduct	s-corti...	lasm
d-seminal vesicle	l-spermatheca	t-yolk	
e-ejaculatory duct	m-spermathecal gland	v-nuc...	
f-ejaculatory sac	n-accessory glands		
g-aedeagus	o-vagina		
h-germ cells	p-micropyle		

The eggs are produced in the females' ov... ...5.2), which are composed of multiple ovarioles connected to the ... ...n oviduct by the calyx. As eggs pass through the common oviduct, ...the spermatheca and the spermathecal gland, which store and nouri... ...after the female has mated. Below the connection to the spermatheca are the connections to the paired accessory glands, which provide secretions that enclose the egg or help to attach it to a substrate. The eggs are laid with the aid of the ovipositor or pass out through the vagina.

The insect egg, or ovum, (fig. 5.3) is formed in the follicle or tube of the ovariole. The yolk material is provided by transport through the epithelial lining of the follicle or in several complex ways by the breakdown of other germ cells or related cells. The chorion, or shell, is formed by the follicular epithelium as the egg passes down the ovariole. The vitelline membrane is formed by the egg cell itself and is essentially the cell membrane. Fertilization is usually accomplished by a sperm cell passing into the egg through an opening in the chorion called the micropyle. This takes place while the egg is in the oviduct or, rarely, in the follicle. The resulting zygote nucleus is surrounded by cytoplasm. The strands of cytoplasm that extend into the yolk are referred to as the cytoplasmic reticulum, while the cytoplasm that surrounds the yolk beneath the vitelline membrane is called the cortical cytoplasm.

In addition to the chorion and vitelline membrane, the egg, when laid, is usually covered by an accessory secretory layer laid on by the follicular epithelium. In some eggs, especially those that must survive winter or pass through a long dry season before hatching, secondary layers beneath the chorion are formed. These additional layers serve to protect the embryo from desiccation.

The embryo begins development when the zygote nucleus begins to divide in a process known as early cleavage. In insects, cleavage usually does not involve the complete division of the egg cell, except in those forms with very small amounts of yolk. In the development of the eggs of Collembola, early division is complete and a solid ball of discrete cells, or morula, is formed, but this then loses its internal organization as the cells fuse to become a syncytium, in which the nuclei and cytoplasm of all the cells are contained within a single membrane. Following this stage of development, the embryo is organized much as in other insects. Similar development also occurs in some minute parasites (Hymenoptera), in which the egg contains very little yolk.

In the higher insects, the large amount of yolk prevents complete cleavage of the egg, and instead most of the cleavage nuclei migrate from the center of the yolk to the cortical cytoplasm to form the blastoderm, or embryonic skin. At this stage, nuclei are not contained in cells with cell membranes about them, but are more or less free. As the embryo differentiates, nuclei concentrate on the ventral surface of the egg, forming the germ band.

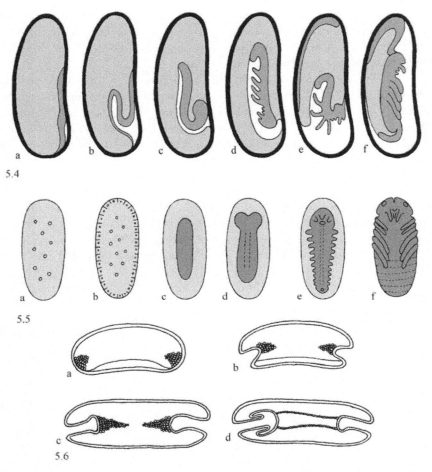

5.4 Development and katatrepsis of embryo. a-formation of germ band, b-anatrepsis of germ band into egg, c-formation of protocephalon, d-condensation of forming mouthparts and limbs, e-katatrepsis of embryo, f-dorsal closure of embryo around yolk.

5.5 Development without katatrepsis. a-early cleavage, b-blastoderm, c-formation of germ band, d-early embryo growing on yolk, e-condensation of embyo showing formation of mouthparts and limbs, f-continued growth of embryo, filling egg and enclosing yolk.

5.6 Origin of the digestive tract in an insect embryo. a-formation of mesoderm, b-infolding of embryo, beginning gut development, c-further infolding, forming stomodeum and proctodeum, d-formation of malpighian tubules and midgut.

As the embryo develops, it performs a number of movements that are collectively called blastokinesis. Depending on the order of insect, this occurs in two different ways. In the Odonata, most Orthoptera, and the hemipteroid insects, the germ band migrates into the yolk by a process called anatrepsis, forming an amniotic cavity. The last part of the germ band to move into the yolk forms the protocephalon, or early head (fig. 5.4). The embryo starts to condense in places and the precursors of the antennae, mouthparts, and legs become noticeable as lobe-like structures. The embryo then moves so that the appendages now face the venter of the egg. This movement is called katatrepsis. The benefit of this movement is that it begins the process of enclosing the yolk inside the embryo. The embryo now grows dorsally to enclose the yolk in a process called dorsal closure. The second form of development is seen in the Dermaptera, Isoptera, Blattodea, Mantodea, and the higher orders. This type of development does not involve katatrepsis; instead, the embryo develops on top of the yolk, which it eventually envelops through dorsal closure (fig. 5.5).

The entire surface of the embryo is the ectoderm, and outgrowths of the wall become the mouthparts and legs. Infolding (fig. 5.6), or an inward growth of cells, occurs at the anterior of the embryo, forming the foregut. An infolding of the posterior of the embryo becomes the hindgut. Outgrowths from the hindgut become the Malpighian tubules. The endoderm inside the embryo differentiates into the midgut, completing the alimentary canal. The nervous system arises from the ectoderm, which forms neuroblasts that differentiate into ganglia and nerve cords. Tracheae may develop as invaginations of the ectoderm, as do the ducts of the reproductive system.

Metamorphosis

Some insects, upon emerging from the egg, appear similar to the adults. Many, however, undergo a variable period of postembryonic development, which may involve very complex changes in body form and function. In the higher forms, this metamorphosis is particularly striking. Embryonic development takes place during the egg stage. After the insect has hatched, entering the larval stage, energy is obtained from a food source, which is often different from the food source of the adult. This energy is used to build up a mass of protoplasm and other materials, which will be converted into the complex adult structures. During the pupal stage (sometimes called the "resting stage" because the insect does not feed and usually is immobile), larval structures are reorganized by complex processes, forming adult structures such as wings, legs, antennae, and reproductive organs. Finally, the definitive molt occurs, beginning the adult stage, during which the insect usually reproduces and dissemination takes place.

These elaborate changes in the life cycle allow the higher insects to build up complex structures with a minimal expenditure of energy. Metamorphosis is also economical in the sense that great numbers of eggs can be produced with a small expenditure of energy per egg. These eggs give rise to relatively simple larvae in large numbers, such that survival is sometimes ensured by sheer abundance. Indeed, massive reproduction seems to be the only real means of survival for some insects.

As mentioned earlier, adults frequently have different food habits from the larvae. For example, butterflies are nectar feeders as adults, but their larvae are herbivores, detritus feeders, or even predators. Similarly, mosquitoes feed on nectar or blood as adults, but their aquatic larvae are scavengers, detritus feeders, algal feeders, or predators. This difference in feeding habits has enabled the evolution of the larvae and adults to proceed independently. Therefore, we often find larvae and adults with such different habits and appearances that they can only be identified as the same species by rearing the larvae to maturity.

Not all insects undergo such profound transformations as flies, butterflies, and beetles. There are four general types of metamorphosis, and of these, the simplest is ametabolous development, in which the insect hatches from the egg as a nymph and undergoes gradual changes through a series of molts until sexual maturity is reached (figs. 5.7 and 5.8). The insect then continues to molt following sexual maturity. The changes from one stage to another are mainly in size, the number of antennal segments, the addition of scales or other integumentary structures, and the maturation of the sexual organs. The noninsectan hexapods (Protura, Collembola, and Diplura) and the apterygote insects (Thysanura and Archaeognatha) are ametabolous.

Paurometabolus development involves the insect hatching as a usually terrestrial nymph without gills (figs. 5.8 and 5.9). The nymph undergoes a series of molts, during which adult characters are gradually added. Sexual maturity is finally reached following a definitive adult molt. After this adult molt, the insect does not molt again. Most of the orthopteroid and all the hemipteroid insects are paurometabolous.

In hemimetabolous development, the insect hatches from the egg as a naiad, which usually has gills and is always aquatic or lives in very wet situations (figs. 5.11 and 5.12). The changes between the immature and adult stages are more profound than in the paurometabolous insects. Mayflies (Ephemeroptera), stoneflies (Plecoptera), and dragonflies and damselflies (Odonata) are the only survivors of what must have once been a very large group of hemimetabolous insects. The mayflies are particularly unique because they emerge from the water with a thin cuticular covering over the body and wings, which must be molted to uncover the genitalia and other body structures. The mayflies are the only order in which a winged insect molts.

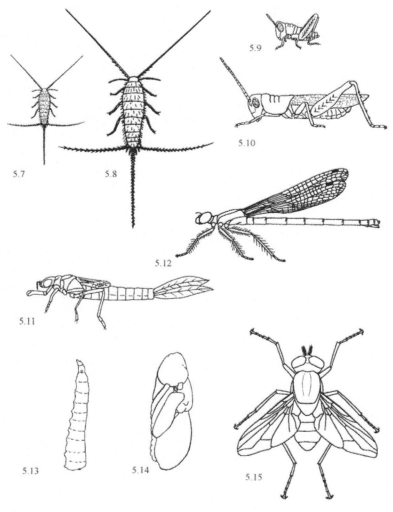

5.7 Thysanuran nymph (ametabolous).
5.8 Thysanuran adult (ametabolous).
5.9 Grasshopper nymph (paurometabolous).
5.10 Grasshopper adult (paurometabolous).
5.11 Damselfly naiad (hemimetabolous).
5.12 Damselfly adult (hemimetabolous).
5.13 Housefly larva (holometabolous).
5.14 Housefly pupa (holometabolous).
5.15 Housefly adult (holometabolous).

Many texts place the paurometabolou[s] hemimetabolous
development into a single category called hemi[metabolo]us or, generally,
incomplete development. These texts basically [make] this a semantic
difference. However, the two forms of developm[ent are i]n many cases so
different that we have chosen to keep this distinctio[n.]

In holometabolous development, the insec[t hatches] from the egg as
a larva, which is usually very different in appearance [from the] adult (figs. 5.13,
5.14, and 5.15). The development usually involves [a pupa, o]r resting stage, as
previously described. Within the pupa, the cells an[d tissues] often break down
so completely that a syncytial condition similar to t[hat in th]e early embryo is
produced. Within this mass, however, various a[reas reta]in the ability to
induce the development of adult structures. Thes[e are calle]d imaginal disks
and represent small groups of cells or nuclei, whi[ch reo]rganize the adult
structure. These disks are already present in the lar[va before] pupation.

Insect Hormones and Moltin[g]

Neurosecretory substances and the produ[cts of v]arious endocrine
glands have a profound effect on the life processes [of insects]. Hormones are
known to influence growth and maturation, molt[i]ng behavior, egg
production, diapause, voltinism (the number of [generation]s produced per
year), control of color, and many other physiologica[l process]es, but it was the
discovery of the hormonal control of insect devel[opment] that has had the
greatest impact on our understanding of the regulat[ion of ins]ect growth.

Three major areas are responsible for [the pro]duction of the
hormones that control molting. The brain has s[everal p]aired groups of
neurosecretory cells that produce neuropeptides, in[cluding pr]othoracicotropic
hormone (PTTH). The prothoracic glands, which a[re located] in the thorax or
head, produce ecdysone (E), a hormone that is r[esponsib]le for the actual
molting of the exoskeleton. The corpora allata ar[e located b]ehind the brain,
and they secrete juvenile hormones (JH), hormones [that con]trol the stage of
development of the molt.

The control of molting involves the timing [and con]centration of the
various hormones (fig. 5.16). First, the brain produ[ces PTT]H, which in turn
triggers the prothoracic gland to release E and th[e corpora] allata to secrete
JH. Early in insect development, the amount of [JH is hig]h. As the molts
continue, the concentration of JH declines. Whe[n JH lev]els are high, the
larval insect molts into another larval stage. Whe[n JH con]centrations have
declined to a certain level, the larva will molt into [the pupal] stage, and when
E is released and JH is absent, the pupa will molt in[to the adu]lt.

It is known from the study of giant chro[mosomes] found in some
insects that these hormones cause "puffs" to appea[r on the] chromosomes at
particular locations. These puffs represent genes [that are b]eing transcribed

into proteins, which will function as enzymes that carry out the necessary activities for molting. Thus, our understanding of the function of hormones in controlling cellular behavior came from the study of insect hormones.

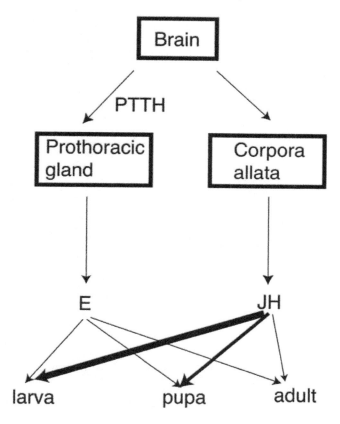

5.16 Hormonal control of insect development.

    Other hormones have been found to play a role in the development process. Eclosion hormone and bursicon are both produced by the neurosecretory cells of the brain just prior to ecdysis. Eclosion hormone triggers the needed muscle movements to actually shed the old exoskeleton, whereas bursicon begins the sclerotization process after ecdysis.
    Many other physiological activities of insects are now known to be directly or indirectly influenced by hormones. Some grasshoppers are differently colored when they develop in different habitats. For example, some species are green in wet meadows and brown if they develop in dry areas. The differences are due to pigments deposited in the cuticle. The deposition of the pigment is under the control of a hormone produced in the

corpora allata. Apparently, the amount of hormone influences the color because the transplantation of extra corpora allata tissue into brown nymphs induces their transformation into green nymphs at the next molt.

Many insects have diapauses or resting stages in their life cycles, which are usually adaptations to allow the animal to survive over cold or dry periods. Diapause is broken by exposure to changes in temperature or some other environmental shock and is also influenced by hormones.

The silkworm, *Bombyx mori*, has several genetic races that differ in the number of generations per year. Some are univoltine, others bivoltine, and some multivoltine. It has been shown that a hormone produced by the suboesophageal ganglion seems to influence the production of diapausing eggs in the univoltine and bivoltine races. Removal of the suboesophageal ganglion causes the production of non-diapausing eggs that will hatch and produce insects that show the same voltinism as their grandparents.

Color in some insects is produced by the presence of pigment granules in the epithelial cells underlying the cuticle and not by pigment in the cuticle itself. The walking stick *Dixippus morosus* (Phasmida) can change its color to correspond with the color of the background or with changes in illumination. In the dark, brown pigment particles migrate to the outer periphery of the cells and the animal takes on a darker color, while in the daytime the brown pigment concentrates at the inner parts of the cells and the insect is lighter. This physiological change is apparently influenced by a hormone from the brain.

Pheromones

Pheromones are mainly substances that influence behavior in much the same way as hormones affect cells, but as their origins are external rather than internal, they may be loosely referred to as ectohormones. Sexual attractants are probably the most common pheromones. These are highly volatile substances released by the females of a species. Males may be attracted from great distances; in some species, they may be drawn from over a mile away. Experiments with the peach tree borer, *Sanninoidea exitiosa* (Lepidoptera), have shown that pheromones may become a possible control for this pest in peach orchards. The males, which normally emerge before the females in the spring, were lured into traps baited with females that had been artificially reared in a lab. By the time the wild females appeared, the male population had been so decimated that relatively few fertile eggs were laid.

In the migratory locust, *Schistocerca gregaria*, sexual maturation is hastened by the presence of mature males of the species. This has been shown to be due to a volatile substance released by males and to a lesser extent by females.

In the honey bee, *Apis mellifera*, the so-called Queen substance, which attracts the worker bees to the queen and is eaten by them, has been extracted and shown to consist of a specific chemical, 9-oxydec-2-enoic acid. It can be placed on a cotton wad, and the worker bees will react to it much as if a queen were present. It is interesting to note that the Queen substance is similar to the 10-hydroxydec-2-enoic acid found in the royal jelly used in rearing queen bees in the hive. Similar pheromones occur in other bees, wasps, and ants, but they are species-specific. Still other pheromones aid in the cross-recognition of members of the same colony and allow intruders to be singled out.

In the termites, the social colonies show a remarkable differentiation of the members into specialized forms called castes. For example, in *Reticulitermes flavipes*, the common eastern United States termite, the colonies contain a queen, a king, secondary reproductives or alates (which are destined to leave the hive after maturity to establish new colonies), soldiers with greatly developed mandibles, and a large number of workers and nymphs. The latter two groups perform most of the work of the colony, such as chewing up wood, building tunnels, and taking care of the queen and her eggs. This remarkable caste structure is apparently controlled by pheromones that circulate through the colony due to constant grooming. If the queen is removed, one of the subreproductive females becomes mature and replaces her. Similarly, the king will be replaced by a subreproductive male. If all subreproductives are removed, soldier-caste nymphs replace them. If all the soldiers are removed, nymphs quickly mature to replace them. In this case, it seems that a pheromone passed from termite to termite in feeding controls the number of each caste that develops, regulating the population structure of the termitarium.

CHAPTER 6: INSECT BEHAVIOR

The study of insect behavior is the examination of their activities as individuals. It follows the review of the central nervous system and the sense organs because these are the structures that provide the insect information about its surroundings and coordinate its various responses. Insect behavior is always purposive, but not necessarily goal-oriented, as most conscious behavior is in humans. Therefore, insect behavior must not be described in an anthropomorphic manner.

There are two basic kinds of behavior: innate behavior and learned behavior. Innate behavior is a more or less fixed response to given stimuli or patterns of stimuli. These inherited responses are relatively inflexible, but can be somewhat modified by learning. Learned behavior, on the other hand, includes responses acquired by an individual through experience and is not inherited.

Some of the innate behavioral patterns in insects include reflexes, such as phasic and tonic reflexes. The former are rapid, short-lived reactions, such as mouthpart activities, while the latter include slower, long-lived reactions, such as posture and body turgor. Most insect reflex patterns involve both phasic and tonic reflexes.

Complex innate behavior in insects is shown in orientation. Primary orientation is the normal body position when the insect is either moving or standing still. Secondary orientation involves changes to the primary orientation when the insect is responding to certain stimuli, such as seeking prey or avoiding a predator. Insects also exhibit kineses and taxes. Kineses are the immediate reactions of the insect to some stimulus. They will result in body movement, but not in the orientation of the body towards or away from the stimulus. Taxes, on the other hand, are behaviors that will orient the body towards or away from a particular stimulus. Examples of taxes include phototaxis (activated by light), hygrotaxis (activated by water), and chemotaxis (activated by odors).

Learning in insects appears to be limited, but some insects can indeed learn. Cockroaches will learn to turn in order to avoid an electrical shock, and bees can learn to discriminate colors if presented with a food reward.

Food-Seeking Behavior

Seeking food is the primary survival activity, given the necessity of obtaining energy for growth and maturation. All insects seek food immediately upon hatching from the egg or being "born" in live-bearing forms, unless, of course, they have been placed directly in or on a food source by their female parent. Food seeking involves a complex of taxes and overlaps with finding a habitat. Oftentimes, this involves some specific stimuli. For example, scarab beetles, blow flies, and other insects feeding or breeding in dung or decaying bodies show a positive reaction toward the odors of skatols, indols, and ammonia, which are normally products of decay. March flies react positively to chemical stimuli, such as the breakdown products of decaying vegetation or hydrocarbons. Mosquitoes react positively to the odors of carbon dioxide, amino acids, steroids, and other volatile substances, and probably to thermal stimuli, but some species feed on cold-blooded reptiles or amphibians. Horse flies react positively to movement and heat (photo- and thermotaxes) and probably ultimately to odor, as they usually do not try to extract blood from the hood of an SUV.

Insects are also attracted to some plants, and the evolution of insects with plants is one of the major developments in the evolution of life on Earth. The flowering plants are closely associated with the evolution of the major orders of insects, particularly the Coleoptera, Lepidoptera, and Hymenoptera. Flowers did not appear until about 125 million years ago. At this time, the major pollinators were beetles. Magnolias and a number of primitive plants with flowers of this type are still beetle-pollinated. Such flowers are composed of numerous spirally arranged petals, stamens, and seed-bearing organs. The petals are usually white or lightly colored, making them stand out against the darker foliage, and they produce large quantities of pollen. The development of brighter colors such as blues, yellows, and reds is probably associated with the diversification of lepidopterans and hymenopterans, as are the deeper, tubular corollas of more advanced flowering plants. Blue and yellow flowers are especially associated with bees, which can differentiate these colors, but are poor at detecting reds. Tubular corollas, which force the insect to penetrate the flower with a long tongue of some sort, allow reduction in pollen production because pollination is more easily assured than in an open flower such as the magnolia. Some plants have become modified so that only a single insect or a few related forms can pollinate them. For example, only certain sphinx moths with enormously long tongues can reach the nectaries of some orchids and, in the process,

provide cross-pollination. Flowers with a distincti...
lip to serve as a "landing platform," occur in many...
and have evolved specifically to be pollinated by be...
evolved more rapidly than those pollinated by bee...
insects, because any change in color, form, size of...
other features by which the bees recognize a hone...
perpetuated as new species. Conversely, the p...
pollinated flowers have remained relatively stable fo...

An example of some of the complexity of...
habitats is shown by an analysis of the behavior...
Pediculis humanus Linnaeus. This louse, in one of...
from the skin and normally clings to the clothing. T...
more frequently on a smooth surface, and turns and...
a rough surface. It is thus more apt to find and st...
also behaves similarly toward a temperature gradie...
body lice are placed on a surface with a heat gradient...
turning more or less at random and congregate...
about 26-30° C (near the normal temperature of th...
diseases such as typhoid fever, in which the tempe...
may be elevated, the body lice may leave the sick...
danger of spreading the infection. A similar seri...
body lice to concentrate where the relative humidit...
normal condition of the air between the human bod...

including a long
groups of plants
lowers may have
other unspecified
dor of nectar, or
would tend to be
"open," beetle-
eriod of time.
food and finding
man body louse,
, feeds on blood
turns and moves
ess frequently on
suitable host. It
t if a number of
ll be moving and
e temperature is
skin). In some
the human body
al, increasing the
ctions will cause
d 76%, near the
thing.

Defensive Behavior

The defense of insects against predators a...
degree against environmental extremes is l...
morphological adaptations or physiological systen...
involve behavior only in their ultimate use. The sti...
is their principal defense, involves behavior in use
defensive chemicals involves behavior.

Many defensive reactions, however, are the...
inducing a specific pattern of behavior. Some ex...
feint and avoidance reactions. The death feint...
mechanism of many insects. When disturbed, the...
appendages and "plays dead." If it is on a plant, it...
and lies there, or it may take off and fly before it h...
feint may last from a few seconds to several minute...

Avoidance reactions are usually triggered...
such as a passing shadow or the sudden removal...
vibration, touch, or sound. The compound eyes...
stemmata are very sensitive to changes in light int...
movement very readily without forming a sharp

ites and to some
ccomplished by
ave evolved and
sp or bee, which
ly, the release of

roduct of stimuli
nclude the death
acteristic escape
ply contracts all
lls to the ground
und. The death

change in light,
low, or by odor,
oup of ocelli or
l probably detect
The larvae of

butterflies and moths show a variety of simple avoidance reactions. When touched, most of them undergo a tensing of muscles and may writhe violently—a common reaction to parasitoid insects. A female ichneumon wasp and a lepidopteran larva may have a violent "wrestling match" while the wasp is trying to insert her egg into the caterpillar's body.

Simple writhing or crawling away may be accomplished in caterpillars by other characteristic reactions. Some, like the larvae of the swallowtail butterfly (Papilionidae), may extrude stink organs. Others rear up and expose eye-like spots that may be frightening to birds or small mammals. Still others, such as the sphinx moth caterpillars (Sphingidae), swing an anal "horn" about as if it were a saber. These reactions include the use of morphological features that have evolved as protective coloration or structure.

Adult moths are especially subject to predation by birds and bats when flying at dusk or at night. Under these conditions, no protective coloration is effective because the insect appears as a black silhouette against the lighter sky, especially on moonlit nights. Bats are the most important predators on moths flying at night, and they have evolved a system for locating flying objects of the right size by sending out extremely high-pitched sounds and listening for echoes (echolocation). This system is similar to the sonar method of locating underwater objects, such as submarines, by sending out sounds and studying the returning sound waves or echoes. However, every weapon the bats have evolved has been countered by an evolving defense mechanism. Many night-flying moths have evolved sonokineses and sonotaxes that help them escape the attacks of bats. On the thorax or abdomen, they have an "ear" with chordotonal organs attached to an "eardrum" or membrane. When this organ receives sound waves in the high-pitched bat range, the moth instantly reacts by going into a death feint-like reaction, usually dropping to the ground or erratically changing its course and flying in a completely random pattern. The bats have countered by changing the aiming point and apparently are successful enough that they make a living. Conversely, the moths escape often enough that, combined with their high reproductive potential, some manage to reproduce and maintain the species.

Reproductive Behavior

Reproductive behavior in insects is extremely varied. It ranges from simple, random male-female contacts, copulation, separation, and egg laying to complex social organizations. Reproductive behavior may be viewed from various standpoints, including morphological, genetic, physiological, behavioral, ecological, or evolutionary. Ecologists may consider reproductive behavior as a system that increases the information in ecosystems and maintains populations in existence under varied conditions. Geneticists may view it as a system ensuring or preventing recombination of genetic materials, and evolutionary biologists may perceive it as a system involving speciation.

Mating behavior usually involves a complete pattern in which activities take place sequentially and one activity may stimulate another. However, we can divide insect reproductive behavior into several phases for discussion: precopulatory behavior, which may include mate finding, mate recognition, display, male rivalry, and sound or pheromone production; copulatory or inseminational behavior; and postcopulatory behavior, including egg laying, male-female cooperation, and nest building.

Mate finding and mate recognition may be entirely random in some insects, but it is suspected that pheromones or other stimuli are involved even in the simplest cases. Some flies seem to attempt copulation with any fly of approximately the same size. Others, however, demonstrate precopulatory displays and postcopulatory protection of the female during oviposition.

Female cockroaches release a pheromone, which acts as an attractant and a powerful stimulus to copulation for the male. In the presence of the pheromone and the absence of females, male cockroaches will attempt to copulate with other males or nymphs. This is also an exception to the rule that pheromones are species-specific, because the female sex pheromone will attract and stimulate males of other species and even other genera.

Many crickets and other insects have precopulatory displays that we often call "mating dances." These may involve the display of wings, special organs, legs, or other body parts, and they may also involve the release of sex pheromones. In some crickets, there are colored glands on the dorsum beneath the wings. Stimulated by the presence of a female, the male may raise the wings, displaying these structures. The female is attracted by them and may begin to feed on the glands and their exudates. This brings her into a position to aid copulation.

In brentid beetles and some scarabs, male rivalry may occur with males "fighting over the female." The conflicts are usually more symbolic than injurious in these heavily armored creatures. In the brentids, postcopulatory behavior may also include cooperation between the sexes, such as the male helping the female extract her ovipositor from the site where she has laid her eggs, thus preventing her from becoming stuck.

Copulation or insemination in insects is usually accomplished by the transfer of a spermatophore, or sperm packet, and usually involves a male intromittent organ. In the silverfish, however, a spermatophore is deposited on a surface near some upright surface or object (the corner of a room, for example). When a female approaches, he "courts" her and more or less guides her toward the spermatophore by spinning silken threads at angles around it.

The male intromittent organs may be elaborate. In fact, in many cases only a very small part of the structure we call the "penis" or aedeagus ever enters the female reproductive tract, but the whole structure is a guide for transferring a spermatophore. The differences between the male intromittent organs and behavior in closely related species may help ensure that there is reproductive isolation between the species. The latter behavior is suggestive of the behavior of flies of the family Empidae (Diptera), which present small insects covered with salivary bubbles (or merely bubbles) to the female before mating. The empids, in fact, show a series of stages in precopulatory behavior, depending upon the size of the male in relation to the female. In genera in which the sexes are about the same size, the male may catch another small insect and present it to the female. He then copulates while she is devouring the prey. In other genera in which the males are smaller than the female, the male first coats the prey with a layer of salivary bubbles which the female must burst before she can eat. In still other genera with very small males, the male does not attempt to catch anything, but simply presents the female with a mass of bubbles, which she proceeds to burst while he is copulating.

Postcopulatory behavior may be relatively simple or very complex. Males often stay with females while they are ovipositing, and a high degree of male-female cooperation may also occur. In order to survive, insects must find a suitable habitat and establish themselves in that habitat. Many insects are helped by their female parent or by both parents in this task. Parental care is known to occur in insects and exhibits a wide range of variation. In some insects, such as walking sticks, care may be limited to simply dropping eggs in a suitable location, while egg brooding is demonstrated by other insect parents, such as some earwigs. Still other insects, such as some wasps and bees, construct special structures for the housing of their young and provision them with food.

Communication in Insects

Some sort of communication is necessary in all dioecious organisms. Males must be able to recognize females, or vice versa, because interspecies matings are generally non-productive. Male butterflies recognize the proper females in some cases by their size, color, shape, and behavior, but the ultimate distinctions are made on the basis of odors, or pheromones. Many grasshoppers, crickets, cicadas, beetles, and other insects communicate by means of sounds. Usually the males produce sounds by stridulation, rubbing a sound-producing structure with another part of the body. This either attracts the proper females or allows the identification of the proper males by the females. The mating swarms of mayflies are also a form of communication, in that a male swarm behaving in a particular way, under the right conditions of temperature, humidity, and light, attracts the proper females. The mating swarms of mosquitoes, such as *Anopheles*, are particularly specific, and in large part prevent the interbreeding of different species.

More complex intraspecific communication is seen in courtship, defense of territory, oviposition, and other behavior, as seen in dragonflies and other insects. In general, however, intraspecific communication is confined to mating behavior except in a relatively few insects.

The content of a communication may vary. In the more social insects, communication becomes very elaborate and not only has a high information content, but makes use of symbols as well as odors, sounds, and physical stimulation. Communication in relation to defense is particularly important in social insects. Termites, for example, apparently not only release pheromones to warn of invasion of the nest and mobilization of the soldiers, but may also beat their mandibles against the substrate and thus synchronize the activities of the whole colony.

In the social bees, communication becomes very complex and involves nest building, nest site finding, and food gathering. All of this complex behavior, however, can be ultimately explained on the basis of rather simple taxes, kineses, and transverse orientations, plus sounds and odors. Many of the steps in the development of the communication in bees such as *Apis mellifera* can be traced among their more primitive relatives and other insects.

One of the primary orienting behaviors in bees is based upon the light compass reaction, which was first studied in ants. At night, most ants use odor trails to guide other ants to food sources or back to the nest. In some, however, the polarization of the light in the blue sky during the daytime is the primary guide. An ant may leave the nest and travel over a very circuitous course for some distance, but on returning to the nest, it may move in a direct line. This use of a sun compass is also characteristic of bees, and a "bee line" is the direct course of bees returning from a honey source to the nest using the light compass reaction. Apparently, the compound eyes adjust the central nervous system to following a particular polarization pattern, which is centered on the position of the sun, even though the latter may be obscured by clouds. This has been shown by experiments with mirrors which reverse the pattern and result in the insects turning and moving in the opposite direction until the normal pattern is again in view.

In *Apis mellifera*, the bees communicate information about honey sources and nesting sites with elaborate dances. These were first studied by Karl von Frisch in Germany, although they had been observed and recorded by Aristotle. The bees perform their so-called dances on the honeycomb, which in this species is hung in a vertical position. This complex behavior requires the transformation of coordinates from the horizontal to the vertical plane and transverse orientations to gravity.

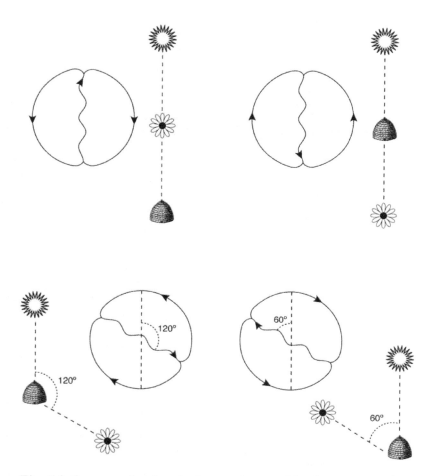

Fig 6.1 Communication in honey bees with the waggle dance. The flower marks the food source in relation to the hive and sun. Diagram shows the waggle dance conveying directional information for the food source in relation to the sun and hive for various situations.

The worker honey bee goes through a series of stages in her ontogeny, which culminate around the twentieth day when she leaves the hive on orientation flights and then becomes a forager for honey. When a forager returns to the hive after having found a nectar source, she usually feeds herself and may feed other bees, and then she performs a series of rapid movements or dances on the honey comb. The round dance is performed when the food source is near to the hive. During the dance, the foraging bee moves in a narrow circle first and then quickly changes direction. When completed, she shares the nectar with the bees that are around her. The other foraging bees then leave the hive and fly nearby, probably detecting the food source by scent. A more elaborate dance is the wag[...] [...] (fig. 6.1), which is danced when the nectar source is some distance [...] hive. It not only communicates the presence of a nectar source and [...] ness through the vigor of the dancing, but also the direction in wh[...] and its distance from the hive. The information is imparted with s[...] accuracy to allow the majority of foragers acting upon it to arrive in [...] ity of the nectar source. Direction is indicated by the angle of the li[...] waggling portion of the dance. Distance is indicated by the duratio[...] dance: the longer the dance the farther away is the nectar source. If [...] source is 100 m away, she will do the waggle dance nine to ten time[...] 5 seconds. If the nectar source is 10,000 m away, she will only dance [...] y 15 seconds.

CHAPTER 7: INSECT EVOLUTION

As noted in the first chapter, there are more insect species in the world than any other group of organisms. In order to understand how this came to be, we must examine the evolutionary process and discuss how it applies to insects.

Our modern view of evolution came as a direct result of Charles Darwin's monumental work published in 1859. *On the Origin of Species* set the scientific world on end by describing how new species could arise from preexisting species, but that concept alone is not why Darwin's work was so influential. His thesis was especially important because it showed that evolution had occurred, and it also described an understandable mechanism that would account for the adaptations of living organisms. This mechanism was natural selection.

Darwin's thesis, now called evolution, can be summarized in a series of points. First, all life tries to reproduce to a maximum. This means that individuals will reproduce more offspring than can easily survive. Second, no two individuals are exactly alike, and most of these variations are inherited. We now know that this inheritance is controlled by the laws of genetics. Third, Darwin noted that there are limits to population growth. There can be only limited numbers of mates and only so much food, water, and space. Fourth, these limits, combined with inherited variations and the fact that all species attempt to produce as many offspring as possible, result in a struggle for existence. Those individuals possessing variations that enable them to be more successful in attracting mates and competing for food will survive longer and reproduce more frequently, leaving more offspring. This differential reproduction is what Darwin called "natural selection." Finally, as populations in different parts of the species' range change in response to different conditions, they will begin to diverge genetically, and eventually may no longer interbreed if they should again come in contact. This is the origin of new species.

Darwin was vexed as to why species are reproductively isolated from each other in this manner. We now believe that this isolation evolved in order to protect the adaptations that have resulted from the evolutionary process. Without genetic isolation, these variations could be lost as species interbred with other species. This gives us the foundation to understand why so many different species of insects have evolved.

Insects played an important role in Darwin's thinking because they illustrate every point of the evolutionary process. Anybody who has ever witnessed an emergence of the 17-year cicadas knows that these insects definitely produce more individuals than can survive. The millions of cicadas that sing in parts of the eastern United States every 17 years and the piles of dead cicadas with deformed wings that accumulate at the bases of trees show that Darwin's first, third, and fourth points are quite true.

Darwin's second point is illustrated by simply examining individual insects belonging to the same species and finding slight differences among them. For example, individuals from a single population of the tiger beetle, *Cicindela repanda*, were found to have elytra that ranged from 6.43 to 8.06 mm in length and 4.35 to 5.57 mm in width (fig. 7.1). Insects also demonstrate that they will change as conditions change. Tiger beetles of the species *Cicindela tranquebarica* become progressively smaller as they range southwestward from New Jersey to Tennessee. This gradual change in a population along a geographic distribution is called a cline. Clinal variations illustrate Darwin's fifth point: physical divergence will occur if populations exist in areas with different conditions. Darwin's use of clinal variation was a masterstroke in his argument, enabling him to demonstrate that species change as conditions change over their distribution. He extrapolated that if this was the case, then species should also exhibit changes if conditions have changed over time.

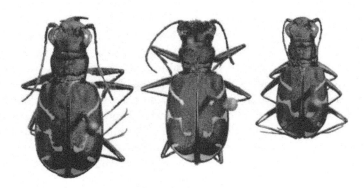

7.1 Clinal variation in *Cicindela tranquebarica* from New Jersey (left), Virginia (center), and Tennessee (right).

Darwin's theory of natural selection was crucial in that it explained adaptation, which can be defined as the "fit" between a structure and its function. Prior to Darwin, it was accepted that adaptations were the result of intelligent design. However, Darwin challenged this idea, citing the example of rudimentary structures. If wings were designed for flight, he noted, then why do flightless birds have wings? If design fails as an explanation for one structure, then it fails as an explanation for all structures. Conversely, the special adaptations found in insects, such as cryptic coloration, made more sense in the light of natural selection.

One of the most remarkable cases of rapid evolution involving cryptic coloration has been discovered among the moths in the industrial midlands of England. Here, where tree trunks are almost always black with soot, melanistic (dark) forms of a number of species of moths have largely replaced the normal lighter or mottled forms of the same species. Experimental evidence has shown that the darker moths have a distinct advantage over the mottled forms when exposed to predation by birds on a sooty tree trunk. Conversely, the mottled form has the advantage when on a normal trunk covered with similarly colored lichens. The change has apparently taken place in less than 125 years and is continuing. While there is still some controversy as to what is the selection agent, there is no doubt of the changing color of these moths during the 20th century.

Some insects are not cryptically colored at all, but rather are so brightly colored that attention is immediately called to them in almost any situation. In many cases, these brightly colored forms are found to be highly distasteful or to possess dangerous poisons or other properties. The exceptions often represent a false warning coloration, by which a tasty or harmless insect gains an advantage by masquerading as a distasteful or harmful one. A well-known example is the case of two common North American butterflies: the viceroy and the monarch. The monarch butterfly larvae feed upon milkweeds and the adults have a nauseous taste or odor, which makes them unacceptable to birds. They are distinctly marked with orange, black, and white, and their behavior indicates that they are not normally eaten by birds. If attacked, they simply go into a death feint and fly off when the attack is over. The larvae of viceroy butterflies feed on willows and the adults are acceptable to birds as food. The coloration of the adults, however, is very similar to that of the monarch. It has been shown that birds that were first fed viceroys would eat them readily, but if afterwards fed a monarch, the birds would then reject both monarchs and viceroys. Birds that were first fed monarchs and then exposed to viceroys later refused both species.

This phenomenon of a harmless or edible insect looking like a harmful or distasteful one is known as mimicry. It is particularly noticeable among tropical butterflies and moths but also occurs in beetles, grasshoppers, wasps, and many other groups. Experiments indicate that as long as the distasteful form (the model) is more abundant than the form that looks like it (the mimic), neither form will be regularly attacked. If the mimic exceeds the model in numbers, however, both forms may be attacked. The advantage to the mimic is therefore relative. Experiments using tasty insects painted with splashes of bright color have documented some natural selection advantage for the insects with the splash of color, and the more similar the insect is to a distasteful one, the greater the selective advantage.

The three types of mimicry are Batesian mimicry, Mullerian mimicry, and aggressive mimicry. Batesian mimicry (so called after H.W. Bates, who first observed it among butterflies in the Amazon region) is the type outlined above in the case of the viceroy and monarch, in which a normally harmless or edible species resembles a harmful or inedible species to which it is not closely related.

Mullerian mimicry, named after Fritz Muller, who first distinguished it from Batesian mimicry, involves a mimetic resemblance between or among a group of distasteful, poisonous, or dangerous forms that are not all closely related. In Mullerian mimicry, all forms are thus equally protected from predators, but a common pattern or resemblance probably has an advantage in that it allows for more rapid indoctrination of young predators. An example of Mullerian mimicry is the close resemblance of many South American butterflies of the families Heliconidae and Ithominidae, in which groups of distasteful species resemble each other closely, although they belong to different families or genera. Mullerian mimicry may also involve Batesian mimicry, in that harmless or edible species may mimic the distasteful forms, and the two types therefore overlap.

Some insects mimic dangerous or poisonous insects, but since they are also predators, the resemblance to a non-predator may increase their advantage over prey species. This is known as aggressive mimicry. For example, the bumble bees (Apidae) seem to have few natural enemies and are mimicked by a considerable number of flies, moths, and other insects. Among these bumble bee mimics are species of predatory flies (Asilidae), which behave like bumblebees and probably gain an advantage in attacking their prey.

Some insects are so peculiarly colored, with great eyespots on the wings or other parts of the body, that it seems probable that they use these spots to frighten off birds or other predators. The behavior of predators in response to such colored insects supports this theory. A. D. Blest has demonstrated that colored eyespots actually frighten birds. He showed peacocks butterflies that have large eyespots on the wings, and when the butterflies opened their wings to reveal these spots, the birds jumped back in fright. More formal experiments using lighted patterns (in place of the butterflies themselves) showed that the reaction of birds became increasingly dramatic as the patterns changed from a plus sign to a single circle to concentric circles to a pair of concentric circles shaded to represent eyespots. Birds often reacted so violently to the latter that they took flight.

Entomologists have had many other opportunities to observe natural selection at work. The most obvious has been the rapid evolution of resistance to pesticides. After DDT was developed, it was widely proclaimed that the house fly would soon be extinct. DDT was a powerful selection agent. It killed billions of house flies over the course of several years. However, not all house flies died as soon as they were exposed. This variation in response to DDT exposure was controlled by an enzyme and was therefore genetically inherited. Those flies not immediately dying had time to lay eggs, while the more susceptible flies died too quickly to reproduce. Soon, most flies had the ability to tolerate some DDT exposure. The degree of this tolerance also varied, with some flies living longer after exposure than others and therefore producing more offspring. Eventually, house flies all over the world were capable of surviving DDT exposure. Rather than succumb to extinction, the house flies had become resistant to DDT, and new and more powerful ways of controlling their populations had to be found. While not a benefit for us, this evolution was a benefit to the house fly and a classic example of natural selection.

The Early Evolution of Insects

The insects have a long evolutionary history that predates the dinosaurs. Though there is still debate as to the origin of the insects, molecular evidence suggests that the insects evolved from crustacea that moved onto dry land. There are fossils of a panhexapod from the early Devonian with a head, thorax, and numerous abdominal segments, with one pair of antennae, three pairs of functional legs on the thorax, and reduced abdominal appendages. These animals are not considered the ancestors of the insects because they were marine and have more than 12 abdominal segments. Rather, they are considered to be members of a larger group of ancient hexapods that includes the insects and their relatives.

Insects have been preserved as fossils in concretions, entombed in amber (fossilized tree sap), as impressions in sandstone and shale, and from tar pits. Evidence of insect feeding can be seen in fossil leaves and in petrified wood. Indirect evidence of insect/plant coevolution can be seen in coal ball peels of upper Carboniferous plant fossils. These fossils provide details of the cell structures of long-extinct plants. One extinct plant group, the seed ferns, produced pollen grains that were large and would have required an insect to aid in pollination, something that is very common today.

The oldest insect fossil occurs at the start of the Devonian period: it is a pair of dicondylic mandibles, but without more fossil evidence, it is not

possible to place this fossil into a particular ins   ler. The oldest
identifiable insect fossil dates from the middle Dev   d is of an extinct
bristletail. These primitively wingless insects lived   rld of Coal Age
forests of unusual trees, and they were contemp   of the first land
vertebrates.

The Upper Carboniferous has preserved a   ent insect fauna.
We will briefly discuss the fossil history of each o   ig orders in later
chapters, but the strange diversity of the insects dur   aleozoic presents
us with many insectan examples that confirm Darw   utionary process.
For example, the oldest mayflies (Ephemeroptera   eir paleopterous
wings, first appear in the Upper Carboniferous.   ayflies, however,
look very different from living mayflies, with fore   d wings of equal
size. Since the Paleozoic, mayflies evolved the red   d wings that they
possess today.

The oldest dragonflies also appear during   r Carboniferous,
along with some now-extinct relatives. The extinc   era is a primitive
dragonfly relative from the late Carboniferous that   s paranotal lobes
with venation on the first thoracic segment. Such   are the strongest
fossil evidence that has been found for a proto-wi   e paranotal lobes
occur in a number of extinct orders, which sup   e idea that they
evolved with the early winged insects.

A spectacular order of dragonfly relativ   the Protodonata
(Upper Carboniferous–Triassic). These insects lo   y dragonfly-like,
but had a unique wing venation. Their front l   ded forward to
capture prey, and the nymphs were aquatic. Some   of this order had
an incredible wingspan (up to 71cm), making then   est flying insects
that ever lived.

The extinct Palaeodictyoptera, Dicliptera   asecoptera, and
Diaphanopterodea were four related orders that sh   ecialized sucking
mouthparts (fig. 7.2). Many Palaeodictyoptera   Carboniferous–
Permian) had paranotal flaps on the first thor   gment, like the
Geroptera, and cerci that were twice as long as the   heir wings had a
dense venation and the hind wings were variable   Some of these
insects had wingspans of up to 56 mm. The nyn   e terrestrial, with
sucking mouthparts similar to those of the adults.   cliptera have lost
the parnotoal lobes, have evolved reduced hind w   ave lost them all
together, and their wings did not have the der   venation. The
Megasecoptera (Upper Carboniferous–Permian)   y similar to the
Palaeodictyoptera, except that the wings were lon   rrow at the base
and some had a pronotum lined with spines. The D   pterodea (Upper
Carboniferous–Permian) differed from the other   n that they had
evolved a folding system for the wings. All four of   ders went extinct
at the end of the Paleozoic.

The Protorthoptera is an artificial order in that it has been used as a placeholder for many fossil insects. To be included in this order, the insect must have had chewing mouthparts and membranous wings with an expanded anal lobe on the hind wing. Some had long cerci and paranotal lobes as well. However, some of these fossils were likely early relatives of the stoneflies (Plecoptera) and the Grylloblattodea. These ancient grylloblattids possessed wings, which their living descendents lost after the late Paleozoic.

Other extinct orders of the Paleozoic include the Titanoptera (Triassic), which were large insects with raptorial forelegs and elongated mandibles. Large, circular, thickened structures on the wings suggest that they may have been capable of sound production. The Caloneurodea (Upper Carboniferous–Permian) had chewing mouthparts, long antennae, forewings and hindwings of equal size, and cursorial legs.

The Protelytroptera (Permian–Cretaceous) were small insects that resembled beetles, but they were actually more closely related to earwigs. They had chewing mouthparts, short cerci, and forewings that were hardened into elytra. The Glosseyltrodea (Permian–Triassic) were possibly holometabolous. They had chewing mouthparts and membranous wings with a unique venation, having an expanded precostal region of the forewing.

The Miomoptera (Upper Carboniferous–Permian) were small insects with mandibulate mouthparts and small cerci. The fore and hind wings were equal in size with reduced venation. They were common in the Permian, and are the most frequently preserved fossil insects in certain localities.

The cockroaches also made their appearance during the Upper Carboniferous (fig. 7.3). However, these early "roachoids" had long ovipositors, which are no longer present in their modern relatives. The first beetles appeared during the Permian period, and their diversity expanded during the Mesozoic. The early beetles possessed wings that retained the venation pattern of their ancestors. Beetle fossils show that the elytra evolved from these membranous wings as the wing veins thickened and hardened cuticle filled in the spaces between the veins.

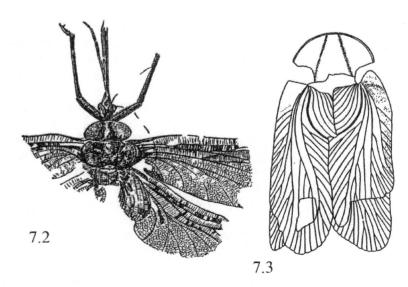

7.2 A fossil Palaeodictyopteran showing the long sucking mouthparts and the small paranotal lobes on the prothorax.
7.3 A Carboniferous fossil roach.

CHAPTER 8: APPLIED INSECT ECOLOGY

Insect ecology is the study of the interactions of insects with both the living and non-living environment. However, insect ecology is fundamentally applied insect evolution, as it deals with three of the five steps of the evolutionary process discussed in the previous chapter: reproductive potential, factors that limit population growth, and the resultant action of natural selection. Applied insect ecology is using this focus to understand how insects become pests, how pesticide applications affect non-target beneficial insects, and how to develop new methods of controlling insect pests.

Applied insect ecology, as used in this text, is also called economic entomology, or in a practical sense, integrated pest management. This emphasis takes into consideration the financial impact of insects in food production, the cost of controlling insects, and the impact that our control measures have on our natural world. In the past, there has been a disconnect between what is necessary for the control of pests and what is best for the environment, but the economic realities of insect control demonstrate that the two concerns are directly connected.

What is an insect pest?

Simply stated, an insect becomes a pest when its populations increase to the point that their activities produce a negative economic impact on our food production, dwellings, or health and well-being. This increase in numbers occurs when the insect's natural enemies no longer keep its population in check. This can occur when a new species is introduced into an area, as in the case of the emerald ash borer that was introduced from Asia, where it is kept in check by other insects.

Insects can also become pests when our activities present opportunities that support an increase in the pest's population. For example, planting a single crop over a large area (monoculture farming) can provide some insects with an essentially unlimited food source, permitting the pest's population to explode.

Occasionally, the environment might change in such a way as to encourage an increase in insect population growth. The European corn borer now can produce two or three generations a year, though it once produced only a single generation per year. This is of major concern as we examine climate change and how it might affect potential pest populations.

The widespread use of pesticides is also creating insect pests, and this phenomenon is best understood when we look at predator/prey models. Generally, the number of prey exceeds the number of predators. If the predators become too numerous, they will overhunt the prey, causing its population to crash. Without other prey for the predators to feed upon, the declining prey numbers will result in a crash in the number of predators. This permits the prey to rebound, which is followed by a subsequent increase in the predators. A classic example of this model was based on observations of snowshoe hare and lynx numbers at Hudson Bay (fig. 8.1).

Insect pests and their natural insect and spider predators show similar population relationships. However, when pesticides are sprayed to control the pest species, the pesticides also kill the predators. Like the hare/lynx relationship, insect pests (the prey) will reproduce faster than their predators, which in many instances leads to subsequent pesticide applications. Each time the pesticide is applied, the predator population is pushed lower, to the point where the predators cannot recover as they would under natural conditions. This will result in a pest population without the natural enemies that could, under normal circumstances, control their numbers. The take-home lesson is that the frequent use of pesticides may, in the long run, create greater pest problems by interfering with the natural balance between prey and predators.

Integrated Pest Management (IPM)

Applying insect ecology to the control of our insect pests grew out of the development and overuse of pesticides following World War II. In times of war, we have always had more casualties from insects than any of our human enemies. Therefore, the development of DDT was of major strategic importance to our war effort. Following the war, the pesticides became available for commercial use in crop production and to promote public health. The result of this widespread use of pesticides was the decline of natural insect enemies and increased pesticide resistance, producing stronger insect pests. As that occurred, it was neccesary to use larger quantities of pesticides to achieve the previous level of control. The fallout of this was increased pesticide residue in the envrionment and the food web, which affected non-target species such as birds. Public concern grew about the effects of pesticides, especially after Rachel Carson's book *Silent Spring* was published in 1962.

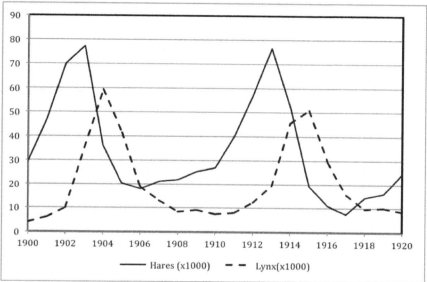

Figure 8.1. A twenty-year sample of Hudson Bay hare/lynx population data. Note the how the predator numbers crash as the prey populations are reduced. Fewer predators permit the prey to rebound. If this were a pest scenario, a pesticide application at the peak of the prey population, when predators are few in number, will drive the predator population further down.

Entomologists responded by applying insect ecology principles for the control of insect pests, and integrated pest management (IPM) was born. IPM decisions approach pest situations in an ecologically and economically feasible way. It encourages environmentally sound decision making, but also recognizes that chemical pesticides have an important role to play in managing pest populations. Before IPM, there were instances when producers and homeowners would spray pesticides on a regular basis even if there was not an obvious pest problem. However, insecticide resistance of the insect pests and the high cost of pesticides provided an incentive to reexamine this practice.

To utilize IPM effectively, the farmer or homeowner must know what pest is present. Insect identification is not just an exercise in understanding diversity. It is the critical first step in IPM decision making. Next, the pest's population size must be determined, which comes from careful sampling of insect populations. The sampling technique used depends on the insect in question. If we were studying the insects on alfalfa, for example, an insect net would be used to sample i[ ]ts in the field. If cockroaches were the target, bait traps would be us[ ]f the methods of collecting insects described in Appendix 1 have [ ]ns in measuring insect populations. Such sampling will be useful i[ ].ining if the pest numbers will result in an economic loss if nothing is [ ]

The critical concepts in the IPM decision[ ] process are the economic injury level (EIL) and the economic thres[ ]). The economic injury level is the point at which the pest popula[ ] will result in an economic loss. When pest populations reach the [ ]ing can be done to avoid some economic loss. To prevent that [ ], the economic threshold (ET) must be determined. The ET is [ ]ulation size that triggers IPM action to prevent the EIL from being [ ]igs 8.2-8.5).

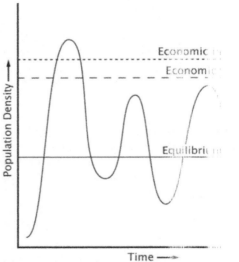

Figure 8.2. The equilbrium position (EP) is th[ ]e population of a pest. In reality, the pest population fluctuat[ ]d the EP. The economic threshold is the pest population [ ]t triggers IPM actions in time to stop any economic loss. If t[ ]mic injury level is reached, the pest will cause an economic loss [ ]roducer.

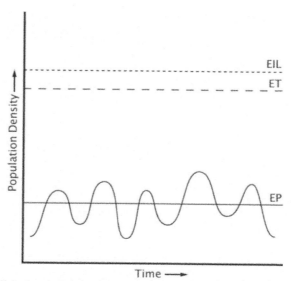

Figure 8.3. This hypothetical population never reaches the ET, so there is no reason to apply a pesticide.

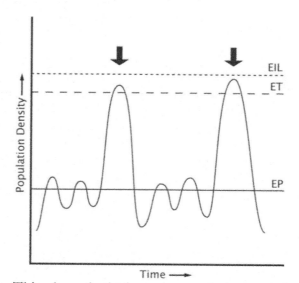

Figure 8.4. This hypothetical pest population reaches the ET, triggering the decision to apply a pesticide (arrows), which successfully lowers the population below the ET. Such a scenario requires constant sampling to monitor the pest numbers.

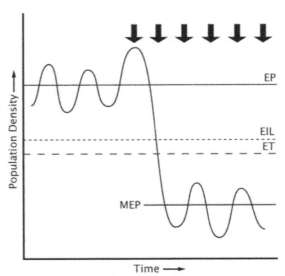

Figure 8.5. In this hypothetical situation, the EP of the pest starts out above the EIL. Therefore, pesticides are applied on a regular basis to force the pest below the ET. The modified equilibrium position (MEP) keeps the pest population in check. Such a scenario would be used when even small numbers of pests may produce cosmetic damage that can impact sales.

Life Tables in IPM

Life tables are important ecological tools for entomologists studying pests or their targets. They provide specific data on the life history, population growth, and interactions of a species, which can be used to determine the EP, ET, and EIL. A typical life table is shown in Table 8.1. This life table, for the fruit tree roller, shows the causes of mortality at each stage of the pest's life cycle.

Life tables can also be developed for a pest's target, such as a food plant, and converted to show the value of the losses at each stage of the plant's growth. Table 8.2, which is similar to table 8.1 in layout, documents the causes of mortality of cabbage plants during a particular growing season. Since cabbage has a market value, the life table can be extrapolated (Table 8.3) to show the economic impact of each of the mortality factors on cabbage that can be used for IPM decision making. If we assume that the cost of applying a pesticide to this cabbage field was $40 per acre, then the application of a pesticide to control cutworms would result in a positive economic return to the producer, because cutworm damage is resulting in nearly $165 of damage per acre. The ET would be the number of cutworms in the field that would produce $40 of damage—the cost of the pesticide. If the cutworm samples exceeded this number, the application of the pesticide would save the farmer the value of the cabbage minus the cost of the application.

The cabbage scenario illustrates the value of using life tables in decision making. For example, if a farmer only had flea beetles feeding on the cabbage, he or she would lose money spraying to control the small number of beetles. In this case, doing nothing would be the best approach. This cost benefit/risk analysis documents the role that pesticides play in IPM. They should be considered the "silver bullet" when quick insect control is critical to maximize profits.

Table 8.1. Life table for Fruit Tree Leaf Roller *Archips argyrospilus* on Apple

Growth period, x	Mean number living per 100 leaf clusters, l_x	Mortality factor, d_xF	Mean number dying, d_x	Percent mortality, $100r_x$
Egg	97	Desiccation	8	8
		Physiological	2	2
Larva, 1-2	83	Dispersion	57	69
Larva, 3-5	26	Parasites	1.0	4
		Predators	0.7	3
		Birds	19.0	73
Pupa	5.3	Parasites	1.0	19
		Predators	0.3	28
		Physiological	0.2	6
		Birds	1.5	4
Moth	2.3	Migration	0.4	17

Generation survival, 1.9-1.96%; mortality 95.1-98.04%.
Modified from LeRoux, E. J. 1963. Population dynamics of agricultural and forest insect pests. Memoirs of the Entomological Society of Canada 32: 103 pp.

Table 8.2. Life table for cabbage, Ottawa, 1968

Growth period x	Mean number living per plot, lx	Mortality factor dxF	Mean number dying per plot, dx	Percent mortality, $100rx$
Establishment	319 ± 4.2	Drought	7.2 ± 1.0	2.2
		Cutworms	56 ± 7.1	17.6
		Root maggot	1.5 ± 0.4	0.5
		Other factors	0.8 ± 0.3	0.3
		Total	65.6 ± 7.1	20.6
Preheading	253.6 ± 6.9	Cutworms	8.6 ± 1.1	2.7
		Root maggot	9.6 ± 1.7	3.0
		Flea beetles	0.3 ± 0.1	0.1
		Rodents	1.3 ± 1.0	0.4
		Clubroot	1.3 ± 0.8	0.4
		Other factors	0.9 ± 0.2	0.3
		Total	22.0 ± 4.5	6.9
Heading	231.6 ± 7.1	Cabbage caterpillars	29.4 ± 1.0	9.2
		Root maggot	0.6 ± 0.1	0.2
		Clubroot	3.2 ± 1.6	1.0
		Soft rot	0.4 ± 0.1	10.5
		Total	33.6 ± 1.7	10.5
Harvest	198.0 ± 6.7	Cabbage caterpillars	18.4 ± 2.2	5.8
		Clubroot	5.5 ± 1.8	1.7
		Soft rot	3.0 ± 0.5	0.9
		Total	26.9 ± 2.1	8.4
Yield	171.1 ± 7.1		148.1 ± 4.5	46.4

Modified from Harcourt, D.G. 1970. Crop life tables as a pest management tool. Canadian Entomologist 102(8): 950–955.

Table 8.3. Operating expenses for cabbage, Ottawa, 1968

Growth period	Potential revenue ($/acre)	Hazard	Loss of revenue ($/acre)
Establishment	$813.96	Drought	$18.36
		Cutworms	$143.05
		Root maggot	$3.83
		Other factors	$2.04
		Total	$167.28
Preheading	$646.68	Cutworms	$21.93
		Root maggot	$24.48
		Flea beetles	$0.76
		Rodents	$3.32
		Clubroot	$3.32
		Other factors	$2.29
		Total	$56.10
Heading	$590.58	Cabbage caterpillars	$74.97
		Root maggot	$1.53
		Clubroot	$8.16
		Soft rot	$1.02
		Total	$85.68
Harvest	$504.90	Cabbage caterpillars	$46.92
		Clubroot	$14.03
		Soft rot	$7.65
		Total	$68.60
Yield	$436.30		

Modified from Harcourt, D.G. 1970. Crop life tables as a pest management tool. Canadian Entomologist 102(8): 950–955.

Proper IPM utilizes a number of strategies to keep pest numbers below the ET, including cultural control, mechanical control, and biological controls. Cultural controls involve methods that make the area less inviting to pests, such as considering the planting site, selecting a cultivar that resists insects, mulching, pruning, watering, fertilizing the soil, or not stacking wood against a house. Mechanical controls physically restrict or remove the pests and may include the use of traps to capture pests, putting netting over trees, and vacuuming or picking off the insects. Biological control employs

practices that encourage the population growth of   enemies, such as
releasing predators to feed on pests, the applicati   cteria and other
pathogens to kill the pests, and the use of pherom   nterfere with the
pest's normal behavior.

The stratagies of IPM do take time to   practice, but the
results can be impressive. A major cemetery and   n in the midwest
has been planting insect-resistant varieties of p   and conducting
proper cultural control for years, and no longer n   icides to control
their plant pests. In addition, a life history study of   oes, which breed
in cemetery-provided flower containers, developed   mendations that
can reduce the disease-spreading mosquito   is with simple
mechanical controls.

CHAPTER 9: INSECTS' PLACE IN NATURE

Because there are more species of insects than any other group of animals, their study requires an understanding of what truly constitutes a species. Carolus Linnaeus, the inventor of the binomial method still used to name species, used a typological concept of species definition, meaning that each species could be represented by a standard or type. Individuals of the species might not be exactly like the type, and Linnaeus allowed for this by recognizing varieties. However, he seems to have been convinced of the fixity of species, and based his work on the idea that there is a recognizable, unchanging type for each species of animal or plant. This is a reflection of the philosophy of his times; in fact, the concept of an ideal type comes directly from Plato.

The present-day concept of a species is biological. Species are now defined as groups of actually or potentially interbreeding natural populations that are reproductively isolated from other such groups. However, there are practical difficulties in applying the biological definition. Most of the insects living today have never been seen alive by the research workers who have named and classified them. Few living insects have been studied genetically, and physiological and ecological studies have been made on only a slightly larger number. Since so much of the information is lacking, we still use the Linnaean methods (without accepting the philosophical basis for them) and base the classification largely upon morphology.

As a consequence, we combine the typological and biological concepts in current taxonomic practice. A species name is based on a type specimen called the holotype, which must be deposited in a museum where it will be preserved. This practice is directly derived from the Linnaean typological concept of the species. A single specimen cannot typify the entire population of a species, but the type system does have its advantages. In species with well-marked morphological characteristics, species names can be more precisely defined by comparison with type specimens than by comparison with published descriptions alone. In many modern descriptions of species, an attempt is made to describe not only the type, but also the range of variation, ecological relations, physiological features, and genetics of the population. Information of this sort is constantly being added to the descriptions of previously established species, though it is not always utilized in classification.

There are, therefore, two meanings of the term "species": morphospecies are recognizable units based on morphological features, whereas biological species are living populations that are reproductively isolated from other populations.

Reproductive Compatibility

All degrees of reproductive compatibility between populations occur, from complete failure to produce offspring (sterility) to complete fertility, as in the population of a species. In some cases, individuals from one end of the range of a wide-ranging species will not interbreed with individuals from the other end, although all the local populations can freely interbreed with those adjacent to them. For instance, the common buckeye butterfly (*Precis lavinia coenia* Heubner) occurs widely over the eastern United States. In Texas and Mexico, it intergrades (or interbreeds) with a closely related but morphologically different population (*Precis lavinia zonalis* Felder) that extends into Central and South America. In that region, the latter population intergrades with other populations and also extends into the Antilles, Cuba, and southern Florida. In Florida and Cuba, the eastern buckeye also occurs, but the two butterflies are not known to interbreed, although they may often be found together on the same flowers.

Reproductive isolation may be due to a number of factors, of which geographic isolation is the most important. Populations isolated on islands, by mountains, by desert areas, or simply by distance undergo evolution independently from the populations from which they were derived. After a period of time, they may be genetically incompatible with their parent species, so that interbreeding will not occur even if conditions change and the populations come back into contact. This is probably the most common way in which species are formed. If all the populations of *Precis lavinia* between Texas and Mexico died out, the two non-interbreeding forms in southern Florida would be considered distinct species. Food plant preferences, habitats, time of emergence, and other factors also influence reproductive isolations in insects.

Isolating mechanisms vary in different groups. Incompatibility may be produced by gene mutations, chromosomal aberrations, or mechanical, physiological, or biochemical incompatibilities. Behavior is one the most common factors preventing interbreeding. In the mosquito *Anopheles maculipennis* complex, for example, the different sibling species do not interbreed because males and females will only mate at specific light intensities and temperatures. Since the different species vary considerably in this regard, males from one species never come into contact with females of another. In many butterflies, males respond sexually only to flying females. They recognize the proper female by size, color, and flight patterns, and ignore female butterflies of other species.

What is a Species?

The preceding discussion of the species concept indicates that the concrete definition of a species is difficult to make. The morphological definition is not satisfactory, and the biological definition is difficult to recognize. Even if we had all the necessary genetic information on all species, we would still not know exactly where to draw the lines. Should two populations be considered distinct if they only interbreed 50% of the time, only 10%, or only 5%? Mathematical calculations indicate that if two populations show as little as a 5% disadvantage in interbreeding, selection under proper conditions will eventually separate them. However, some natural populations have an even greater genetic disadvantage in breeding, but still do not show any sign of separation.

The best definition for our purposes, therefore, is a simple and practical one: an insect species is the smallest group that we can recognize and differentiate from other groups. This definition will work well in any one location, but it cannot be extended to a large area because no two populations are ever genetically identical. In a small area, the local species are usually separated from one another by physical or behavioral differences, so that they represent discrete and definable units. Over large areas, the local species lose their discreteness through intergradation with related populations, adaptations to different conditions, and the accumulation of different genetic mutations.

Higher Categories

The categories above the species level vary in number and usage in different groups. People who work with the classification of animals (taxonomists) are variable in their behavior. Some are called "splitters" and some are "lumpers." Splitters tend to recognize many species, genera, families, and classes, "splitting" groups of insects into as many categories as possible. Lumpers tend to use the fewest possible categories. Excessive splitting or excessive lumping causes confusion and conceals information about the relationship of different groups. Categories such as superfamilies, subfamilies, and tribes are therefore useful to separate larger recognizable groups while keeping closely allied groups or species together.

The Genus

A genus (plural: genera) is composed of closely related species that have descended from a common ancestor. Members of a genus are usually adapted to a single way of life, which is referred to as an ecological niche. For example, all of the species of *Anopheles* are blood-sucking, nectar-feeding mosquitos whose larvae feed at the surface of fresh water. The species of the genus *Cychrus* (Coleoptera: Carabidae) are all predators on land snails in the eastern forests of the United States, and their larvae are predators in the leaf cover of the forest floor.

Ideally, genera are natural units separated by distinct differences from other genera. Some genera, however, are not natural units, and combine species that only look alike because they have converged in evolution. These accidental genera are eliminated as soon as they are recognized.

The Family

The family is a group of related genera, the species of which have descended from a common ancestor. Families show great diversity in the specializations of the genera; the members of some genera of a family are adapted for one way of life, while those of another genus of the same family are differently adapted. For example, all of the genera of mosquitoes (the family Culicidae) are not composed of blood-sucking nectar-feeders. Some feed only on nectar. Similarly, the larvae of different genera do not all feed at the surface of water, but vary in their habits. Some are predators, some feed on detritus on the bottom of a pond or stream, and some feed only on specific diatoms or algae. The habitats of both adults and larvae also vary from genus to genus.

The Order, Class, Phylum, and Domain

Orders are groups of related families. Orders, classes, and phyla are characterized by basic structural patterns that are shared by all their members, but which show special adaptations in the lower categories. Thus, orders are based on differences in wings, mouthparts, appendages, and other body parts. With some exceptions, such as the Siphonaptera, the orders are not based on adaptive features except in the broadest sense. The families, genera, and the species within an order may be adapted to very different ways of life such as predation, parasitism, and scavenging.

Domains are defined by cellular complexities. The domains Bacteria and Archaea are composed of simple cells that do not have their genetic material organized within a true nucleus. The Eukarya, the domain to which all plants, fungi, and animals (including insects) belong, are distinguished by a membrane-enclosed nucleus that protects the chromosomes.

Distribution of Insects

Most of the orders and many families are worldwide in distribution and occur on all the continents except Antarctica. A few orders, however, represent ancient groups that have died out over much of the earth. Among the insects, the Grylloblattodea occur only in western North America and Japan, and the Diploglossata, often included in the Dermaptera, are confined to Africa.

Four orders of insects—the Coleoptera, Lepidoptera, Diptera, and Hymenoptera—predominate in the world today. These orders probably contain about 85% of the living insects. Many other orders are well represented in most areas. Some orders that have few species compared to the major orders are well known because of their habits or behavior. The Thysanura, Blattaria, and Siphonaptera often occur in houses, and the Orthoptera are abundant and their activities often attract human attention.

Entomologists use biogeographic regions to generalize the distribution of an insect. These terms were created by Alfred R. Wallace, who studied the distribution of animals and plants and was a close friend of Charles Darwin. He found that life on Earth is generally distributed into seven geographic regions: Nearctic, Neotropical, Palearctic, Ethiopian, Oriental, Australian, and Oceanic. The Nearctic includes North America from central Mexico northward. The Neotropical region extends from Central Mexico south, including South America. The Palearctic includes Europe and Asia, whereas the Ethiopian includes Africa south of the Sahara. The Oriental region includes China and Southeast Asia, and the Australian region includes Australia and Tasmania. Oceania encompasses the island life of the oceans. These terms are particularly useful in describing the higher taxa, but are often too broad to help with the lower taxa.

The terms used in describing insect distribution in North America are generalized as follows:

Cosmopolitan—occurring all over the Earth, either naturally or having been transported by commerce. The latter category includes many household insects, micropredators such as bed bugs and fleas, and agricultural pests.

Widespread—occurring on both sides of North America and usually extending from our southern to northern extremes.

Eastern—occurring in the United States and Canada east of the Rocky Mountains, but rare or lacking on the west coast.

Western—occurring in the western part of the United States and Canada, but not extending far eastward.

Southern—occurring in the southern United States and eastern Mexico.

Northern—occurring in the northern United States and Canada, but not extending to the Gulf Coast.

Southwestern—occurring in the southwestern U[nited Stat]es and northern Mexico.
Northeastern—occurring in the northeastern Unit[ed States] and Canada, but not extending far to the west or south.

Some other terms used, such as accidental o[r ...] (introduced by commerce), or "tropical, wandering north," are mor[e or less] self-explanatory.

The distribution of many arthropods is m[ore stron]gly influenced by ecological requirements than by strictly biogeogra[phical co]nsiderations. In the stoneflies, for example, the naiads are poorly [adapted] for respiration, and are largely confined to cooler regions where [coo]l streams occur. Other groups, such as the scorpions (Scorpionida), [are serio]usly regulated by temperature relationships, and do not extend fa[r to the] north except as accidentals. In general, however, the distributi[on of a] group is largely influenced by the degree of adaptive radiation tha[t has occu]rred. (Adaptive radiation may be defined as the spreading out and e[volution] of several species to occupy different niches and habitats.) The majo[rity o]f insects contain species adapted to almost every conceivable habitat [and way] of life.

Sizes of Insects and Their Re[lative]s

The size of arthropods varies consider[ably in d]ifferent groups. When describing insects in general terms, the follo[wing def]initions will help keep things in perspective.

Minute—less than 1 mm long. The smallest adul[t arthrop]ods are found in the classes Protura, Collembola, and Insecta. [In th]e latter, some Hymenoptera (*Alaptus*) are only 0.21 mm long, and [some C]oleoptera, some Ptiliidae (*Nanosella* living in pores of mushrooms[) are on]ly .25 mm long. Immature stages, of course, may be extremely small.
Very Small—1 mm to less than 5 mm long.
Small—5-10 mm long, or less than 15 mm a[cross the] wings. Many arthropods collected by the beginner will fall into th[is group.]
Large—20-50 mm long, or 25-75 mm across the w[ings.]
Very large—over 50 mm long, or more than 75 m[m across] the wings. The longest insects known are Phasmida from the Eas[t Indies,] which may reach 325 mm in length, but most of these are very thin [and elon]gate. Among the Lepidoptera, the largest moths may reach a wingsp[an of ...] mm. Terrestrial arthropods other than insects may also be very la[rge, such] as land crabs in Florida and Mexico and crayfish throughout the wo[rld.]

Arthropod Diversity

Insects belong to the phylum Arthropoda, [a large] group of animals

that are found in the fossil record dating back to the beginning of the Paleozoic Era, some 540 million years ago. Most arthropods are readily assigned to classes on the basis of clearly defined external characters, and nearly all can be recognized by some or a combination of the following characters:
1) The body is covered with a tough, non-living cuticle, which is hardened in restricted areas or forms an exoskeleton composed of hardened plates.
2) The body is segmented, with intersegmental grooves between some segments.
3) The body segments are grouped into functional units (tagmata).
4) The paired locomotory appendages are jointed, and are often covered by cylindrical sclerites separated by flexible membranes.
5) Some anterior jointed appendages are modified as mouthparts for feeding or as antennae or other sensory organs.

Among the living arthropods are seven typical body plans, and most of these are highly characteristic of one or more classes or parts of classes. In the Xiphosuran form, seen in the horseshoe crabs, the anterior dorsal sclerites are fused into a shield-like structure. The abdomen is similarly comprised of fused dorsal sclerites and a long terminal spike-like structure called the telson (fig. 9.1). Horseshoe crabs are placed in the Class Xiphosura, a marine group with only a single American species, *Xiphosura polyphemus*, which can reach lengths up to 60 cm long. They range from Maine to Panama.

The Pycnogonid form, or sea spider, has a reduced number of body segments and usually has long legs. The sea spider's body is so small that the digestive and reproductive organs extend into the legs (fig. 9.2). They are all placed into the Class Pycnogonida, and all are marine.

The Arachnid form has the first nine or more segments fused to form a cephalothorax, and the remaining segments form an abdomen, either with segments free or more or less fused (fig. 9.3). Arachnids are a diverse group that includes spiders, harvestmen, and scorpions.

The Acarine form, including the ticks and mites, are arachnids in which the anterior segment and mouthparts form a piercing structure called the gnathosome, while the remainder of the segments are fused into one mass, with segmentation often indicated only by legs and other organs (figs. 9.4 and 9.5).

The Crustacean form has nine or more anterior segments usually fused to form a cephalothorax, a variable number of free abdominal segments, and a terminal telson. The crustacean body plan (fig. 9.6) is often highly modified, as seen in the crabs, lobsters, shrimp, clam-like fairy shrimp, ostracods, barnacles, and many others. There is one terrestrial group of crustaceans, the isopods or pillbugs, which also have a characteristic form (fig. 9.7).

The Myriapodan form (centipedes and millipedes) has several anterior segments more or less fused to form a definite head and the remaining segments free, forming an elongate trunk with a pair of legs on each segment. This body plan is found in the classes Pauropoda (fig. 9.8), Diplopoda (fig. 9.9), Chilopoda (fig. 9.10), and Symphyla (fig. 9.11).

The Hexapod form (insects and insect-like animals) has the first six segments fused to form a definitive head, and the next three segments are often fused or partly fused, forming a thorax that is equipped with three pairs of strong locomotory appendages. Finally, the remaining segments form an abdomen. The hexapod body plan includes the classes Protura (fig. 9.12), Diplura (fig. 9.13), Collembola (fig. 9.14), and Insecta (figs. 9.15 through 9.19).

These various body plans illustrate the adaptive abilities of the phylum Arthropoda. The exoskeleton, combined with articulated limbs, provided arthropods the potential of invading varieties of habitats. The arthropods with mandibles took two routes: one type colonized the oceans and became the crustaceans, and another moved onto land and eventually exploded into the insects. Along the way, as the various body forms illustrate, evolution tried a number of things, but the arthropods as a group are so successful that many of these experimental forms are still with us. For example, there are clearly four living classes with the hexapod body form, but it is in the class Insecta that the arthropods reached their greatest diversity. Their evolutionary success is reflected by their place in the higher classification of arthropods (Tables 9.1 and 9.2).

The higher classification of insects is a matter of intense debate because the classification should reflect our current understanding of the evolution of the insect orders. New molecular procedures and new fossil discoveries are presenting entomologists with a better understanding of the evolutionary relationships of insects. At the time of this writing, there are two differing classifications of the arthropods. The Atelocerata classification combines the myriapod and hexapod body plans into a single taxon, the Atelocerata (Table 9.1). The opposing view combines the crustacea and the hexapods into the Pancrustacea and separates the Myriapoda into their own taxon (Table 9.2). This latter view is gaining support based on new molecular studies that suggest a freshwater crustacean as the hexapod ancestor.

Insects are, however, a single class of arthropods, which does not illustrate the range of diversity that has evolved in the class. It is at the ordinal level of classification that the different adaptations of insects can best be appreciated. As with the higher classification of the arthropods, the relationships of the insectan orders are hotly debated. New molecular analyses are grouping some orders into superorders to reflect their common ancestry. The classification used in this text is presented in Table 9.3.

Table 1
ARTHROPOD CLASSIFICATION
Atelocerata view

Kingdom Animalia
    Phylum Arthropoda
        Subphylum Trilobita (extinct)

        Subphylum Chelicerata
            Class Xiphosura – horseshoe crabs
            Class Pycnogonida – sea spiders
            Class Arachnida
                Order Scorpionida - scorpions
                Order Araneae – spiders
                Order Opiliones – harvestmen
                Order Acari – mites and ticks

        Subphylum Crustacea – lobsters, crabs, shrimps, barnacles

        Subphylum Atelocerata
            Class Diplopoda - millipedes
            Class Chilopoda - centipedes
            Class Pauropoda - pauropods
            Class Symphyla - symphylans
            Class Protura - proturans
            Class Diplura - diplurans
            Class Collembola - springtails
            Class Insecta - insects

Table 2
ARTHROPOD CLASSIFICATION
Crustacean view

Kingdom Animalia
    Phylum Arthropoda
        Subphylum Trilobita (extinct)

        Subphylum Chelicerata
            Class Xiphosura – horseshoe crabs
            Class Pycnogonida – sea spiders
            Class Arachnida
                Order Scorpionida - scorpions
                Order Araneae – spiders
                Order Opiliones – harvestmen
                Order Acari – mites and ticks

        Subphylum Myriapoda
            Class Diplopoda - millipedes
            Class Chilopoda - centipedes
            Class Pauropoda - pauropods
            Class Symphyla - symphylans

        Subphylum Pancrustacea
            Class Crustacea – lobsters, crabs, shrimps, barnacles

        Epiclass Hexapoda
            Class Protura - proturans
            Class Diplura - diplurans
            Class Collembola - springtails
            Class Insecta - insects

Table 3
THE CLASSIFICATION OF LIVING INSECTS

Class INSECTA

    Order Archaeognatha
    Order Thysanura

Infraclass Pterygota

    Order Ephemeroptera
    Order Odonata

Division Neoptera

    Subdivision Polyneoptera
      Order Plecoptera
      Order Embiidina
      Order Phasmida
      Order Orthoptera
      Order Mantophasmatodea
      Order Zoraptera
      Order Blattaria
      Order Isoptera
      Order Mantodea
      Order Dermaptera
      Order Grylloblattodea

    Subdivision Paraneoptera
      Order Psocodea
      Order Thysanoptera
      Order Hemiptera

    Subdivision Endopterygota
      Superorder Neuropterida
        Order Megaloptera
        Order Rhaphidioptera
        Order Neuroptera

        Order Coleoptera
        Order Strepsiptera

      Superorder Hymnopterida
        Order Hymenoptera

      Superorder Panorpida
        Order Mecoptera
        Order Siphonaptera
        Order Diptera
        Order Trichoptera
        Order Lepidoptera

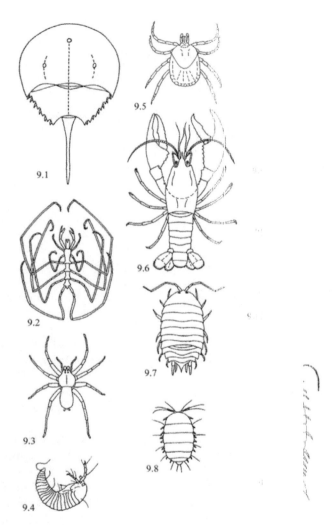

9.1 Xiphosuran form, horseshoe crab.
9.2 Pycnogonid form, sea spider.
9.3 Arachnid form, spider.
9.4 Acarina form, mite.
9.5 Acarina form, dog tick.
9.6 Crustacean form, crayfish.
9.7 Crustacean form, pill bug.
9.8 Myriapodan form, pauropod
9.9 Myriapodan form, millipede.
9.10 Myriapodan form, centipede (Bailey).
9.11 Myriapodan form, symphylan (Hilton).

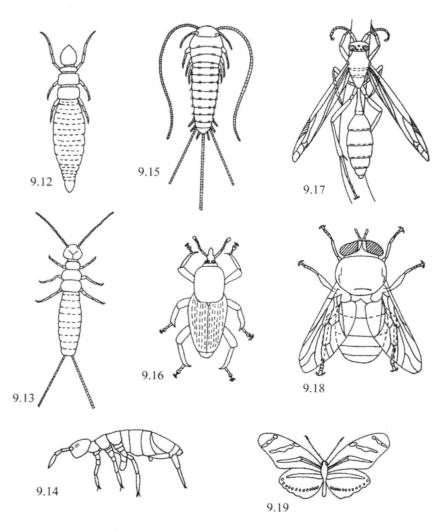

9.12 Class Protura , proturan.
9.13 Class Diplura, dipluran.
9.14 Class Collembola, springtail.
9.15 Class Insecta, silverfish.
9.16 Class Insecta, beetle.
9.17 Class Insecta, wasp.
9.18 Class Insecta, horse fly.
9.19 Class Insecta, butterfly.

CHAPTER 10: ENTOGNATHAN HEXAPODS

The members of the classes Protura, Collembola, and Diplura have, in the past, all been included within the class Insecta. Recently, they have been placed into a monophyletic group: the Entognatha.

The noninsectan hexapods are similar to insects in that they have a head, a thorax with three pairs of legs, and an abdomen, but they can be separated from the insects by their entognathous mouthparts (retracted into the head), thin cuticle, and lack of well-developed compound eyes. The primitive wingless insects, on the other hand, have mandibles that are not retracted into the head, compound eyes, a typical insect cuticle, and no muscles beyond the first segment of the antenna.

Class Protura

The Protura (fig. 10.1) are very small to minute animals that are less than 2 mm long and are usually found in soil or in debris associated with the soil. They are elongate and flexible and lack eyes, antennae, and cerci. The integument is weakly sclerotized, usually nearly colorless, and has relatively few setae. The head is pear-shaped or nearly circular, and sometimes anteriorly pointed. The mouthparts are entognathous, retracting into a pouch within the head capsule when not in use. The anterior legs are often elongate and extend in front of the body, where they are apparently used in place of antennae. The thoracic segments are loosely united with the prothorax and are comparatively small. In some, the thoracic segments are not very different from those of the abdomen, but the thoracic legs are always strongly developed. The abdomen is elongate, starting with nine segments when the proturans hatch from the egg and increasing to twelve segments after three molts in the adults. Metamorphosis is ametabolous with a definitive adult molt, and fertilization is indirect.

The Protura show affinities with the myriapods in the addition of body segments after hatching and in presence of styli (vestigial abdominal legs) on the first three abdominal segments. Most of the external features, such as the flexible body, thin integument, reduced appendages, retractable mouthparts, and reduced body setae, are correlated with life in the soil. They feed on decaying matter.

Only a single order of Proturans is currently recognized. Fossils of these delicate hexapods have not yet been found.

Class Collembola

The Collembola (figs. 10.2–10.7), or springtails, are small to minute animals, with the Nearctic species ranging from less than 1 mm to about 6 mm in length. They are usually found in soil, litter, and seawrack, or under rocks and other covered habitats where they feed on decaying matter. A few may be found on vegetation, but most remain closely associated with the soil. Their integument is thin, but may be colored and covered with setae, flattened scales derived from setae, wart-like structures, or papillae. The antennae are four-segmented, and some segments may be secondarily divided. Simple eyes are often present in groups on either side of the head behind the antennae. The mouthparts are entognathous. The abdomen is six-segmented, although the segments may be largely fused in the adults. A characteristic ventral tube or sucker-like organ called the collophore is always present on the venter of the first abdominal segment, a forked organ called the retinaculum is usually present on the third segment, and a forked springing organ, the furcula, is found on the fourth segment. The collophore is involved with water uptake, and the retinaculum and furcula function in a manner similar to an inverted catapult, propelling the collembolan forward and upward up to 100 mm. Fertilization is indirect and involves a spermatophore. Metamophosis is ametabolous, and adults may molt up to 50 times. The tracheal system, when present, opens through a single pair of spiracles between the head and prothorax.

The Collembola have many specialized setae and sensory organs. The setae may be plumose, clubbed, or modified into flattened scales. Besides the simple eyes, pseudocelli, which resemble tiny eyes, may occur in various parts of the body.

The springtails are sometimes placed in the same class as the Protura, but their collophore, retinaculum, and furcula are unique. These appendages clearly develop from abdominal legs during embryonic development.

The Collembola are among the oldest terrestrial arthropods known. They are first found in the fossil record in the Devonian, along with some of the oldest terrestrial plant fossils.

Class Diplura

The Diplura (figs. 10.8-10.9) are elongate white or translucent animals that may be up to 25 mm long. They are usually found in caves, soil, debris associated with soil, or under roots. All species are eyeless, but the segmented antennae are well developed, and the mouthparts are entognathous. The thoracic segments are loosely united, and the prothorax is small in comparison to the other segments. The thoracic legs are well developed and the tarsi each have two claws. The abdomen is elongate, with ten segments, and is terminated by cerci, which are either elongate with several segments, or one-segmented and modified to form terminal forceps. Fertilization is indirect, with the males producing a spermatophore that is placed at the end of a small stalk. Some dipluran females guard their eggs and young, but risk being eaten by their cannibalistic offspring. Development is ametabolous, and the adults may molt up to 30 times.

Molecular evidence suggests that the Diplura, Collembola, and Protura evolved from a common ancestor. Eyelessness, a flexible body, and the thin integument correlate with an underground habitat. Like the Protura, the diplurans feed on decaying matter.

The Diplura have a long fossil history, with representatives extending back to the Lower Cretaceous and possibly the Upper Carboniferous.

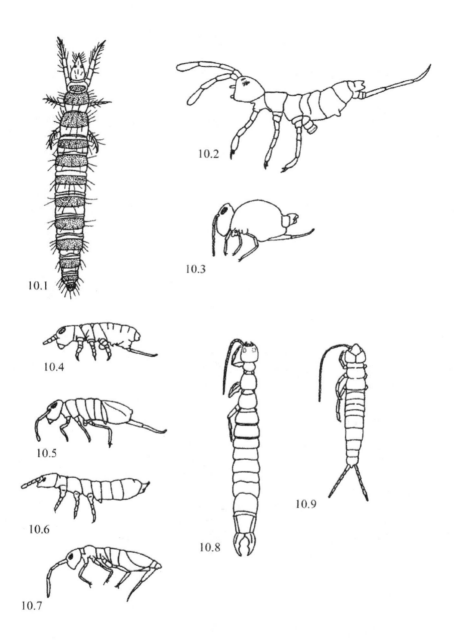

10.1 Proturan adult.
10.2–10.7 Collembolan adults.
10.8–10.9 Dipluran adults.

CHAPTER 11: THE PRIMITIVELY WINGLESS INSECTS

The primitively wingless insects bear the clos... mblance to the hypothetical ancestral insect. They retain, in one ... her of the living forms, a number of physical and developmental fe... it may have been inherited from a crustacean ancestor. These inclu... i or simple eyes, compound eyes, a median abdominal filament prob... ologous with the telson of Crustacea, retention of eleven segments in ... men, a primitive arrangement of the nervous system, retention of ... l spermatophore transfer, and the lack of a definitive adult molt in th... e.

Order Archaeognatha

The Archaeognatha (figs. 11.1 and 11.2)... as the jumping bristletails, range from 7 to 15 mm in length. ... have large eyes, monocondylic mandibles (having a single articulation ... head), (fig. 11.3), large multisegmented maxillary palpi, and long ante... yli (fig. 11.2), the remnants of legs, are found on the abdominal segm... still retain some function in helping to support the abdomen ... walking. The archaeognathans also have two large cerci and ... medial filament. Their bodies are rounded and have an arched or hu... l appearance.

Whereas sperm transfer is external, ferti... ccurs internally. Mating involves a dance or courtship, during which ... e places a sperm packet on a silken thread or stalk and then directs th... over the packet, affecting fertilization. In other archaeoganthans, th... ctually places the sperm packet directly on the female's ovipo... Development is ametabolous. Before successful molting, the insect ... itself to a surface using a cement derived from its feces. If the cemen... not hold firm or the surface is not solid, the insect will not be able to ...

Archaeognathans are found in woods, under logs, and in leaf litter, where they eat plant debris, algae, or lichens. Their common name, jumping bristletails, is deserved as they can jump up to 30 cm.

The Archaeognatha may be represented by fossils dating back to the Devonian, but they definitely occur in the Upper Carboniferous.

Order Thysanura

The Thysanura (fig. 11.4) include the silverfish and firebrats, which range from 2 to 22 mm in length. Their bodies more flattened than in the Archaeognatha and the integument is usually covered with scales. The compound eyes are smaller than those of the Archaeognatha and, in some species, may be altogether absent. The mandibles are dicondylic, meaning they have two articulations with the head (fig. 11.5), which links the Thysanura with the winged insects. The antennae are long and filamentous. The cerci and median filament are also elongate and all three structures, unlike those of the Archeognatha, are approximately the same length.

Sperm transfer is external, and there is a courtship during which the male places a spermatophore on a silken thread or vertical surface and may direct the female to it by spinning additional threads. The female will take up the sperm packet with her ovipositor, completing fertilization internally. As in the Archeognatha, development is ametabolous and there is no definitive adult molt.

Originally, the thysanurans lived in caves, beneath bark, or in leaf litter, where they functioned as scavengers on decaying matter. However, some species of thysanurans have become pests in homes and libraries due to their tendency to eat starchy materials such as bookbinding glue and paper.

Definitive fossil representatives of the Thysanura extend back to the Cretaceous.

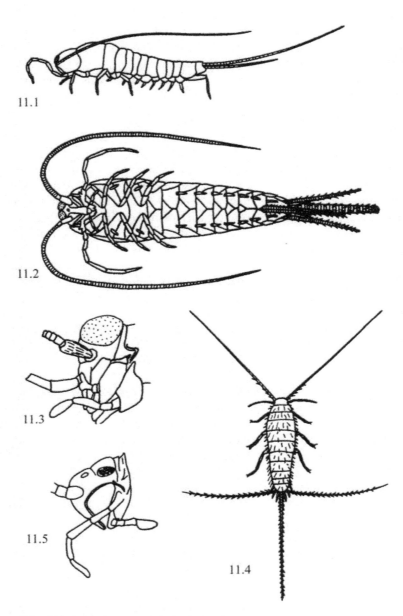

11.1 *Machilis* (Archaeognatha), lateral aspect.
11.2 *Machilis* (Archaeognatha), ventral aspect (showing styli).
11.3 Lateral aspect of head of *Machilis* (Archaeognatha).
11.4 *Termobia* (Thysanura), dorsal aspect.
11.5 *Tricholepidion* (Thysanura), lateral aspect of head.

CHAPTER 12: THE PRIMITIVELY WINGED INSECTS

All the winged insects belong to the infraclass Pterygota, but the orders Odonata and Ephemeroptera differ from other winged insects in that they cannot fold the wings flat over the back when at rest, but hold them straight to the sides, slanting backward, or upright over the back. This paleopterous condition is also found among some of the extinct orders including Palaeodictyoptera, Megasecoptera, and Protodonata (Chapter 7).

The Odonata and Ephemeroptera resemble each other in having membranous wings with many longitudinal veins and a rich network of cross-veins called the archedictyon. They are hemimetabolous, with the aquatic immatures or naiads respiring by means of gills. The naiads are usually equipped with well-developed compound eyes and ocelli.

Order Ephemeroptera

Adult mayflies (fig. 12.1) are delicate insects with membranous wings in which the axillary sclerites are well developed. The wings do not fold flat over the back, but are usually held vertically and closely together above the back when at rest. The mouthparts of the adults are non-functional, and the digestive system has been modified into an air sac or aerostatic organ, so that the adults are specialized for mating and dispersal and live only a short time.

The adult mayflies retain a number of primitive characters, such as the elaborate network of veins (figs. 12.3) and elongated many-segmented cerci. Some adults also possess a median filament probably retained from an ancestral telson. In other respects, they are more specialized than the Odonata. The legs are slender and adapted for grasping, but are sometimes greatly reduced. The hind wings are sometimes lacking (fig. 12.2) and in all cases are smaller than the forewings, allowing better coordination in flight. In some Old World genera, the wing venation is strong but reduced, and these forms are as adept fliers as the dragonflies.

The male genitalia of the mayflies are primitive but well developed. They are composed of a pair of claspers and a double aedeagus between them. Each half of the aedeagus connects directly with the primary gonoduct of its respective side. The female's two genital openings are behind the seventh sternite, and each connects with the ovary of its respective side. Mating takes place in flight, with the males forming characteristic mating swarms. The swarms move up and down in graceful rhythmic patterns, and females, apparently attracted by the swarms, are clasped by the males from below and copulation takes place with the male in the inferior position.

All adult mayflies have well-developed compound eyes. In some species, the males have the eyes divided into an upper supposition eye and a lower apposition eye, which is thought to aid in spotting mates in the mating swarms.

The Ephemeroptera are hemimetabolous, and the naiads can molt as many as 35 times prior to molting into the winged form. A special adaptation is the presence of a thin, waterproof cuticle that covers the newly molted adult. This covering is usually shed after emergence from the water has been accomplished. This is the only order in which a winged adult molts, and this special winged form is called a subimago.

The naiads are aquatic and respire by means of gills, which are leaf-like, branched, plumose, segmented, or otherwise modified. The gills are homologous with legs, developing in the embryos from leg rudiments, and they usually occur only at the sides of the abdominal segments. The naiads occur in fresh water or, rarely, in brackish situations. The body form often reflects the habitat in which the naiad is normally found, with burrowing, rock-crawling, swimming, clinging, and scrambling types easily recognized. The greatest diversity of species is found in streams in temperate regions or at moderate elevations in the tropics. Most late naiads have a multi-segmented median appendage as well as elongate lateral cerci on the abdomen (figs. 12.4 and 12.5), and this readily distinguishes them from the naiads of the Plecoptera, which have only the lateral cerci developed as filaments. The naiads feed on small aquatic organisms and debris, whereas the adults do not feed.

The short life span of adults is reflected in the name Ephemeroptera. Some species are extremely fugitive, living only for brief periods around sunset or dawn at very restricted periods of the year.

Mayflies extend in the fossil record back to the Upper Carboniferous. These early mayflies had forewings equal in size to the hind wings, and by the Jurassic mayfly hind wings had reduced in size to the characteristic form seen in figure 12.1. Some mayflies have lost the hind wings entirely, as seen in the family Caenidae (fig. 12.2). The large number of naiad molts and the presence of the subimago support the view that the Ephemeroptera are the most primitive of the extant winged insects.

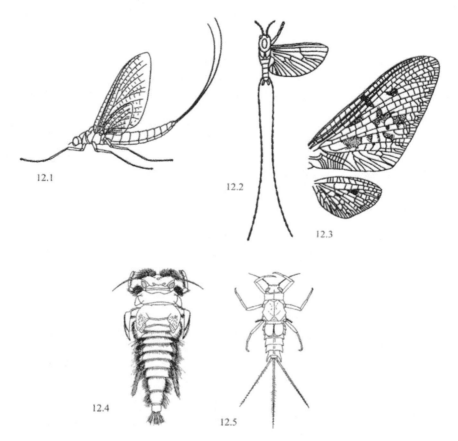

12.1 Mayfly adult, Ephemeroptera.
12.2 Small mayfly adult (Ephemeroptera, Caenidae).
12.3 Fore and hind wings of adult mayfly (*Ephemera*).
12.4 Dorsal aspect of the naiad of *Dolania* (most of the caudal filaments omitted).
12.5 Dorsal aspect of the naiad of *Neoephemera*.

Order Odonata

The adult damselflies (fig. 12.6) and d... ... (fig. 12.7) are medium-sized to very large insects, with some me... ...er 15 cm across the wings. They possess two pairs of membranou... large compound eyes, conspicuous ocelli, strong mandibles, and sp... labial palpi. The sclerites of the body wall are moderately well scler... ...d the wings and body are often brightly colored, especially in m... ...e wings have a characteristic break in the costa, called the nodu... ...ere is usually a colored spot known as the pterostigma near the wi... ...he fore and hind wings are generally similar in shape and venation... hind wings are usually broader and have a greater extension ... ...nal area in the dragonflies. Dragonflies and damselflies cannot f... wings; however, some damselflies have a modification of the thora... ults in the wings being held vertically over the back of the insec... ...h the meso- and metathorax are fused into a large flight box, or pter... the power stroke is generated by muscles attached directly to the s... ... the wing bases. The upstroke is generated by indirect muscles, as i... r winged insects. The legs are slender, adapted for grasping, and arm... numerous spines. The abdomen is usually long, slender, and cylindric... netimes flattened or club-like toward the tip. The antennae are minu... ...e labial palpi are expanded and strongly toothed.

The mating system of Odonata is unique. ... genital opening, as is usual in winged insects, is between the ninth... ...th sternites, but there is a unique copulatory apparatus on the sec... third abdominal segments, which requires that sperm be externally ... ...d from the male genital opening to this copulatory apparatus befor... with the female. The male abdomen terminates in unsegmented cer... ... are adapted for grasping the female behind her head. After she is ... ...sped, the female brings her genital opening, which is between the ... ...d ninth sternites, forward and upward to contact the male copul... ...aratus. In this position, the pair form the characteristic "copulator... which can often be observed around ponds or along streams. Copul... ...en takes place in flight and may be preceded and followed by elab... urtship activities conducted by the males. In some species, the ... t only court the females before copulation, but also defend territori... other males and attend the females during oviposition.

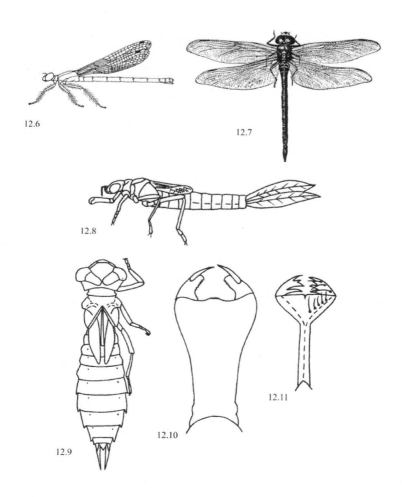

12.6 Damselfly, suborder Zygoptera.
12.7 Dragonfly, suborder Anisoptera
12.8 Typical appearance of a zygopteran nymph (*Ischnura*, Coenagrionidae) with the labial mask extended.
12.9 Typical appearance of an anisopteran nymph (*Basiaeschna*, Aeschnidae).
12.10 Ventral view of the labial mask of an anisopteran nymph (*Boyeria*).
12.11 Ventral view of the labial mask of the zygopteran nymph (*Lestes*).

The Nearctic species have naiads of two general types: the naiads of damselflies (suborder Zygoptera) have leaf-like gills at the tip of the abdomen (fig. 12.8), whereas the naiads of dragonflies (suborder Anisoptera) have internal rectal gills (fig. 12.9). The dragonfly naiads not only bring water into the rectum for respiration, but can also expel it forcefully enough to propel themselves away from predators. All naiads feed by use of the specialized labium (figs. 12.10 and 12.11), which is extensible and forms a mask-like covering over the other mouthparts when in repose. Both the naiads and adults are predators, and development is hemimetabolous, with the naiads molting between 9 and 17 times.

There are two suborders of Odonata in the Nearctic. The Zygoptera are the delicate damselflies (fig. 12.6), and the Anisoptera are the more robust dragonflies (fig. 12.7). Their fossils extend back to the Permian period and show the evolutionary changes in wing venation that produced diagnostic venation characteristic of the order.

CHAPTER 13: THE POLYNEOPTERA

The remaining orders of insects are all grouped within the division Neoptera because they have in common the feature of neoptery: the capability to fold the wings against the body when they are not in use. A few wingless orders are included in the division, but they evolved from winged ancestors. Their placement into the Neoptera and other taxa is based on other shared features.

The Polyneoptera, sometimes referred to as the orthopteroid insects, consists of several orders that have chewing mouthparts, cerci, and many Malpighian tubules. The winged forms have richly-veined wings, but the archedictyon seen in the paleopterous insects is usually reduced or lacking and the hind wing has a large anal lobe. Eleven living orders are included in this group, including many that are commonly encountered in the home and garden. Three of the orders, the Blattaria, Mantodea, and Isoptera, share a common ancestry and are sometimes classified together in the superorder Dictyoptera. The evolutionary relationships of the other orders to each other have not been resolved.

Order Plecoptera

The Plecoptera, or stoneflies, are an ancient order of insects. Like the Odonata and Ephemeroptera, the stoneflies have terrestrial adults (fig. 13.1) and aquatic naiads (fig. 13.2). This similarity in life cycles led to their being grouped together in the past, but they are not closely related. The Plecoptera, being neopterous, are considered to be more advanced than the paleopterous dragonflies and mayflies.

The Plecoptera are small to moderately large insects with two pairs of membranous wings. The sclerites of the body wall are thin and lightly sclerotized. The wings are usually well developed, but brachypterous (shortened) or micropterous (extremely small) in some. The hind wing usually possesses an anal fold at its base (fig. 13.3). The venation is primitive, with retention of the archedictyon in some forms. Compound eyes and two or (usually) three ocelli are present. Antennae are filiform, with numerous segments. Mouthparts are adapted for biting, but mandibles are sometimes vestigial in the adults. The thorax is loosely compacted, with the pro-, meso-, and metathorax distinctly defined. The legs are well developed for running, walking, or climbing, and the abdomen is elongate with ten evident segments and vestiges of an eleventh segment at base. The cerci are usually long, with many segments. Metamorphosis is hemimetabolous, but naiads are similar to the adults except for the lack of wings.

The aquatic naiads (fig. 13.2) usually have tufted tracheal gills on the head, pronotum, or thoracic segments, or on both the thoracic segments and the abdomen. More rarely, anal gills are present in rosettes or patches, along with other gills. The species that lack gills are capable of respiring directly across the body wall. The tracheal gills may rarely be retained in the adults as vestigial structures.

The naiads are usually adapted for scrambling over the stream bottom, where they eat a variety of foods. Most are herbivores, but some are carnivorous. The antennae and cerci are long, with the median filament undeveloped. Compound eyes and ocelli are usually present.

A few naiads are found under moss or in moist debris near water. Most species are found in streams, and only a few species inhabit ponds or slow-moving or stagnant water. The body of stream forms is usually depressed, much as in mayfly naiads, but many that live in moss, gravel, or in organic debris in water are more rounded in body form and show little modification of the legs. A few forms such *Peltoperla* (Perlidae) have enlarged flaps on the thorax, which cover the gills so that these species can live in saturated sediments on the stream bottom.

Adult stoneflies have an unusual method of attracting mates. They use a hardened structure on the venter of the abdomen to tap or rub against a surface. The recipient stonefly responds in a similar method. It is thought that these acoustic signals enable the stoneflies to communicate without giving away their location to potential predators. The male zeroes in on the female by signaling, waiting for an answering signal, and then searching for its source. The process is repeated until they find each other.

Adult stoneflies display a mixture of advanced and primitive characters. The major part of the life cycle is spent in the immature stage; the adults often do not feed and in that respect are similar to the mayflies. In some species in northern regions, the adults appear during the late fall or winter, apparently to escape predation and competition with other insects. These are called winter stoneflies in order to differentiate them from the summer stoneflies, which emerge in the warmer months. The adults of many species are remarkably agile runners, which possibly gives us a living demonstration of what the early neopterous insects were like: sure-footed and swift like their thysanuran ancestors. Adults are detritivores or omnivores, but some carnivory is known. Stoneflies with vestigial mandibles do not feed as adults.

Plecopteran fossils are found dating back to the Permian period, with the oldest forms possessing paranotal flaps and hind wings with large anal lobes. There are two suborders that reflect a northern and southern hemisphere distribution, which supports the hypothesis that these suborders evolved after the supercontinent Pangaea broke apart in the early Mesozoic.

Order Embiidina

The Embiidina, or webspinners, are moderately small gregarious insects ranging from 3 to 20 mm in length (fig. 13.4). Only the males possess two pairs of similar membranous wings, which are shed after the male finds a permanent home. The wings have a greatly thickened radius, whereas other veins are reduced or absent. The wings are usually a smoky color, with clear spaces between the veins that give this insect its characteristic appearance (fig. 13.5). Adults and nymphs have three-segmented tarsi. The first segment of the prothoracic legs is always enlarged and contains silk glands connected to hollow setae (fig. 13.6). The mouthparts are adapted for chewing. The antennae are moderately long, with about 20 segments. The webspinners have two segmented cerci, which are asymmetrical in the male. Metamorphosis is paurometabolous, with the adults differing little from the nymphs.

The embiids live in small to moderately large subsocial colonies in the ground or among debris. Adults and nymphs spin silk, which line their galleries' passageways. Females show considerable parental care for the eggs and nymphs. The embiids feed on decaying plants and lichens. The adult male does not feed, and uses its mandibles to hold the female during mating.

Definitive embiid fossils date to the Cretaceous, but a protoembiid is known from the Jurassic.

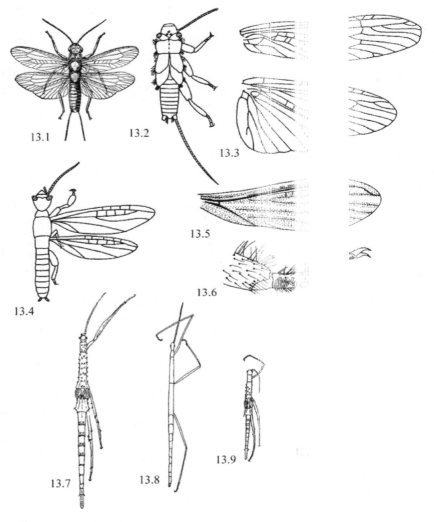

13.1 Plecopteran adult.
13.1 Plecopteran naiad.
13.3 Fore and hind wings of a stonefly. Note the   nf anal lobe.
13.4 Adult male embiid.
13.5 Forewing of embii male.
13.6 Fore tarsus of an embiid showing silk blade
13.7 Walkingstick, *Aplopus*.
13.8 Walkingstick, *Manomera*.
13.9 Walkingstick nymph, Phasmida.
13.10 Walkingstick nymph, Phasmida.

Order Phasmida

The Phasmida (figs. 13.7–10), or walkingsticks, are sometimes very large insects, with the North American species ranging from about 14 mm to over 100 mm in body length. Nearly all, except the Old World tropical leaf insects (Phylliidae), are slender and cylindrical creatures, well deserving of their common name. With the exception of a few species from Mexico and Florida, the North American species are wingless. Phasmids are best represented in the tropics, where many show elaborate thorn-like or leaf-like projections on the body and legs. Some tropical phasmids have the longest body size of all living insects, some reaching lengths of over 325 mm. The head is usually free and moveable, with well-developed compound eyes. The ocelli are often lacking, and the antennae are usually long. The mouthparts are typically adapted for chewing. The prothorax is short or very short, even in elongate forms, while the mesothorax is short or elongate and the metathorax is often very long and closely united to the first abdominal segment, with the suture between them sometimes obscured. The abdomen is usually long, cylindrical or tapering, and the cerci are unsegmented. The legs are nearly always long or very long and similar, with the tarsi five-segmented or rarely three-segmented. Metamorphosis is paurometabolous. The eggs are variable in shape; some resemble seeds and are collected by ants that are attracted to a nutrient knob on the egg. The eggs are laid singly and may take years to hatch. Many eggs are resistant to fire and other forms of damage. When they hatch, an operculum (lid-like structure) opens and the nymph emerges. Some walkingsticks are capable of parthenogenesis and do not produce males. All species are terrestrial and vegetarian in all stages, and one species produces toxic exudates in a gland on the prothorax that serve as a defense against predators.

The phasmids are an ancient and highly specialized group of insects. They are probably an early offshoot of primitive orthopteroids that specialized as camouflaged herbivores, dependent for survival upon their resemblance to the plants on which they fed. Fossil phasmids are known from Baltic and Dominican amber dating back to the Miocene, but fossil phasmid eggs have been found in Cretaceous rocks.

Order Orthoptera

The Orthoptera, a large order with approximately 22,500 species, includes the grasshoppers, katydids, crickets, and their relatives. They range from small to very large insects, 5 mm to over 100 mm long, with saltatorial hind legs that have the femora enlarged for jumping. Some orthopterans also have fossorial forelegs modified for digging or burrowing. The mouthparts, which are usually directed downward or sometimes forward, are adapted for chewing, and the antennae are usually well developed. The forewings, if present, are at least partially thickened into leathery coverings called tegmina, and the hind wings fold in a more or less fanlike manner. Many wingless forms or forms with reduced wings occur. The tarsi may have between one and four or rarely five segments. The prothorax is usually large and sometimes greatly enlarged and freely articulated. The meso- and metathorax are united, forming a pterothorax. Metamorphosis is paurometabolous with the nymphs resembling the adults. The Orthoptera are nearly always terrestrial, but a few species live in damp conditions or are even aquatic, diving into water to escape enemies. Most of them are herbivores, but there are exceptions.

Sound-producing and sound-receiving organs are often present. The grasshoppers produce sounds by rubbing pegs on the inside of their hind legs against the edge of the forewings. Crickets and katydids sing by rubbing a file-like structure on the underside of the wings against the hardened edge of the wing beneath it. The sound-receiving organs in grasshoppers are found on both sides of the first abdominal segment, while the crickets and katydids have their "ears" on the front tibiae (fig. 13.12).

There are two suborders in the Orthoptera that are easily separated by their antennal length. Except for the mole crickets (fig. 13.11), the Ensifera have long antennae, usually with more than 30 segments. The Ensifera include several common families, such as the Gryllidae or true crickets (fig. 13.14), the Gryllotalpidae or mole crickets (fig. 13.11), the Tettigoniidae or katydids (figs. 13.12 and 13.13), and the Rhaphidophoridae or cave and camel crickets (fig. 13.15). Members of the suborder Caelifera have short antennae and include the family Acrididae or grasshoppers (fig. 13.17), and the family Tetrigidae or pygmy grasshoppers (fig. 13.16).

The Orthoptera is an ancient order, with fossils extending back to the Permian. The oldest fossils belong to the Ensifera, with the first Caelifera appearing in the Triassic.

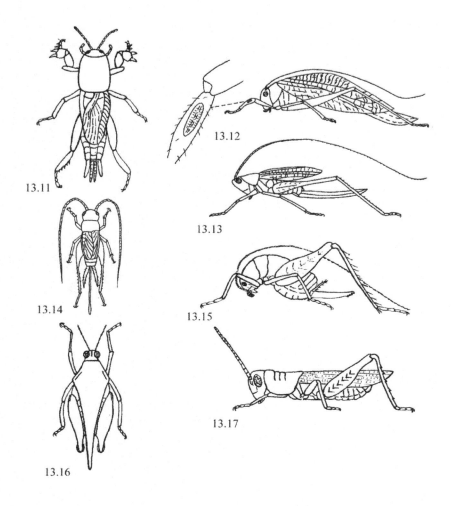

13.11 Mole cricket, Gryllotalpidae.
13.12 Katydid with auditory organ on the fore tibia, Tettigoniidae.
13.13 Female katydid, Tettigoniidae.
13.14 Field cricket, Gryllidae.
13.15 Camel cricket, Rhaphidophoridae.
13.16 Pygmy grasshopper, Tetrigidae.
13.17 Grasshopper, Acrididae.

Order Mantophasmatodea

The Mantophasmatodea (fig. 13.18), described in 2002, was the first new order to be discovered since 1914. They are wingless insects that have chewing, hypognathous (downward-directed) mouthparts, long antennae, cursorial legs, and one-segmented cerci. Most of these insects are nocturnal predators of small insects. They are called heelwalkers because they walk with their last tarsal segments raised, not touching the substrate. They have a paurometabolous life cycle.

Members of this order have been collected for years in Namibia, South Africa, and Tanzania, but they were previously misidentified as nymphal walking sticks. Fossil Mantophasmatodea have also been found in Eocene Baltic amber.

Order Zoraptera

The Zoraptera (fig. 13.19) are small insects with chewing mouthparts, short cerci, two-segmented tarsi, six Malpighian tubules, and only two abdominal ganglia. The apterous forms are common in large populations, but when the colony becomes crowded and resources decline, winged forms are produced that disperse to a new decaying log. Winged zorapterans possess four wings with reduced venation, compound eyes, and three ocelli. The apterous (wingless) forms are blind and lack ocelli. Metamorphosis is paurometabolous.

The zorapterans resemble nymphal wood roaches or termites. They are gregarious, but not truly colonial, and live in decaying wood, where they feed on fungal spores, hyphae, and small arthropods. They can be collected in sawdust piles, often being found on the same substrates as termites.

The Zoraptera are found in the fossil record dating back to the Cretaceous period.

Order Blattaria

The order Blattaria includes the cockroaches, which are among the most common insects encountered by humans. They may be characterized as small to very large insects, ranging in size from 2 mm to over 100 mm in length. The head is characteristically deflected beneath the thorax and concealed by the broad pronotum, so that it is nearly or quite invisible from above. The legs are adapted for running, with five-segmented tarsi. The mouthparts are typically adapted for chewing and include distinct mandibles, maxillae, and a labium. There are usually two ocelli and the compound eyes, when present, are kidney-shaped. The antennae are usually long, often being much longer than the body (fig. 13.20). The pronotum is large and shield-like, and the meso- and metathorax are more or less united to form a pterothorax. The wings are often absent or reduced, but when fully developed, the forewing is tegminous, with the venation partly obscured. The hind wings are usually folded fan-like beneath the forewings and have a distinct anal fold at the base. The forewings, in repose, usually overlap to some degree on the back. The wing venation approaches the hypothetical primitive type and sometimes includes remnants of the primitive archedictyon. Metamorphosis is paurometabolous, with nymphs similar to adults except for the lack of wings in winged forms. The eggs are characteristically enclosed in a purse-like egg case called the oothecum, which may be deposited in a suitable place, carried about by the female, or retained in the female abdomen until the eggs have hatched.

Most cockroaches in temperate regions are rather drab yellowish, brownish, or blackish insects, whereas tropical species show much less restraint. Some of these are pale green, resembling leaves, while others presumably mimic brightly-colored distasteful insects. Among the latter are bright orange-and-black species, which resemble lycid beetles (Coleoptera: Lycidae); brightly metallic species, which resemble wood-boring beetles (Coleoptera: Cerambycidae, Buprestidae); and shining black species that resemble the highly odiferous cydnid bugs (Hemiptera: Cydnidae). Some Neotropical cockroaches roll up into a ball and resemble pill bugs (Crustacea: Isopoda).

The habits of many cockroaches of living in dark, sheltered places, such as under debris or in hollow trees, preadapted them to invade human habitations and other buildings. At least five species are now cosmopolitan in distribution and scavenge on a wide array of foodstuffs.

The majority of modern species are solitary, although some may aggregate due to characteristics of the environment. In one Australian species of *Panesthia*, however, males and females live together with their offspring in burrows in the ground, and *Cryptocercus* is found in the forests of eastern and western North America, where they live in small colonies in dead wood and show a subsocial organization. The latter cockroaches eat dead wood, which they digest with the aid of rich intestinal flora. The flagellate protozoa that assist wood digestion in *Cryptocercus* are similar to those found in some termites (Isoptera).

About fifty species of roaches have invaded our homes and are considered pests. American cockroaches (fig. 13.21) and Oriental cockroaches belong to the family Blattidae, and the German cockroach and the common northern wood roach (fig. 13.22) belong to the Blattellidae. The family Blaberidae includes the pests *Leucophaea* and *Blaberus*.

Cockroach-like fossils are found in the Upper Carboniferous. These early "roachoids" differed from modern roaches in that females had long ovipositors . True roaches evolved by the early Cretaceous period.

Order Isoptera

The order Isoptera includes the termites, [text cut] responsible for considerable structural damage to our homes. [text cut] termites can be characterized as small to medium-sized polymorp[hic] insects living in colonies, usually in soil and with or without an a[bove] ground mound nest. Termite colonies often include reproductive, sub-[reproduc]tive, soldier, and worker castes, which may be further differ[entiated in] size and other characteristics (figs. 13.23–13.27). The wings occu[r in] the reproductive caste and for only a brief period during swarmi[ng and] [fo]rmation of new colonies, after which they are shed along a suture [near their] base (fig. 13.28). When present, the four wings are membranous [and simil]ar in shape and venation. The wing venation is complex, varying f[rom prim]itive to reduced, with the veins rather weakly developed and th[e primit]ive archedictyon retained. The mouthparts are of a simple biting [type, except in certain soldiers, in which the mandibles may be reduced o[r the head is] prolonged into a rostrum. The antennae and palp[s are well] developed in all castes. Compound eyes are also well developed i[n reprodu]ctive castes, but reduced or lacking in others. The ocelli are usuall[y present when compound eyes are present. The tarsi are four-segmented, [and the] genitalia rudimentary or lacking in both se[xes]. [D]evelopment is paurometabolous.

The termites are closely related to the co[ckroaches] but show many parallels with the true ants (Hymenoptera: Formici[dae] [wit]h which they are sometimes confused. Morphologically, termites res[emble co]ckroaches except in the position of the head, which is not deflected b[eneath th]e pronotum, and in the membranous wings, which have not been r[educed i]nto tegmina. In some primitive Australian termites, however, the hi[ndwings] have an anal area which folds fan-like as in the cockroaches, and the [way] the wings, after the membranous portion is shed, resemble small te[gmina. F]urther evidence of relationship to the cockroaches is indicated by [the prese]nce of subsocial behavior in the cockroaches *Panesthis* and *Cryptocer*[cus, the] wood-feeding habits of the latter associated with mutualistic proto[zoa].

Polymorphism in termite colonies is someti[mes extr]eme. Four castes are nearly always recognizable. Reproductives, wh[ich are re]presented by the queen (fig. 13.24) and an associated male, the k[ing, after] characterized by having had wings during the mating flight or swar[m, although only the wing stubs are retained (fig. 13.25). After colony formation, t[he queen] and king raise a brood of young termites, which then take over the [operatio]n of the colony, leaving the reproductives little to do except reprod[uce. In m]any species, the queen rapidly becomes physogastric, with the enla[rged abd]omen containing little except the hypertrophied ovaries and develop[ing eggs]. She is then fed, cleaned, and otherwise tended by the workers. [The kin]g is usually little changed. At specific times of the year, winged se[xual] reproductives are formed; they swarm from the colony, pair with mates, and found new colonies.

Subreproductives (fig. 13.23) are often recognizable in colonies. These are adults that fail to develop wings or develop only rudimentary wings and thus do not swarm. They are capable of becoming reproductive and replacing the queen or king of a colony if the latter are removed.

Soldiers (fig. 13.27) are specialized for protection of the colony from ants and other enemies. They may be differentiated into a number of classes according to size. When an opening is made in a termitarium, soldiers crowd into the breach, and either stop the invaders or fill the breach with their termite dead. The mandibles are often very elaborately modified, but the significance of all of the modifications is not known, except that many seem adapted for attacking other insects. The nasutiform soldiers have their mandibles reduced, but the head is prolonged into a tubular rostrum that serves as a "squirt gun." The large frontal glands discharge an acrid volatile substance, which is squirted upon enemies invading the colony and apparently strongly affects the nervous system of ants and other insects. The secretions of the frontal gland may also be adhesive, so that when squirted upon an insect, they impede movement.

Workers (fig. 13.26) are relatively nymph-like adults that may, like the soldiers, be differentiated according to size. Different size classes may perform different functions within the colony. In general, the workers perform all the functions necessary for maintenance of the termitarium, such as building and cleaning galleries and mounds, feeding the young, caring for the eggs, gathering food, and tending the fungus gardens in the more specialized termites.

The termites represent the only insect order in which all species are social. The social organization is at the level of the more highly evolved ants, bees, and wasps of the order Hymenoptera. In fact, with the possible exception of the ants, the social behavior of termites is among the most highly developed of any animal. Colony formation and integration, parental care of the young, differentiation of castes and division of labor, restriction of reproduction, food storage and processing, and construction of elaborate nests are well developed in all termites.

Except for the subsocial cockroaches, there are no living insects that illustrate the probable course of the evolution of social behavior in the termites. It is clear, however, that the termite colony is a single family produced by a single pair of parents. The cockroach *Cryptocercus* undoubtedly illustrates one of the stages through which the termite society developed. The next step was probably the development of the elaborate food and exudate exchange, which is so characteristic of termites, followed by specialization of the members of the colony for different functions. One of the most important factors in the lengthening of the lives of the reproductive castes is to ensure that they overlap the lives of their offspring. Some termite queens may live up to 50 years. Social organization would be impossible in insects having only a short lifespan that ended before the offspring were mature.

The exchange of food and exudates in the termite colony was called trophallaxis by William Morton Wheeler. This process is the binding force that holds the termites together. There is constant licking of the other individuals, exchange of food by mouth, perianal feeding, and related activities between members of a colony. The queen, and to a lesser extent the king, are constantly groomed by the workers. This even extends to the eggs, which will not hatch unless they are licked and fondled. As they grow, the young nymphs are also licked, fed, and tended by the workers.

Trophallaxis provides a rapid diffusion of food through the colony, as well as of pheromones, which obviously influence all the life processes of the individual. It also furnishes the circulation necessary to distribute the chemical materials controlling maturation, growth, and caste formation. Because this efficient communication enables all members of the colony to act with a single purpose, several researchers have analogized the termitarium to a super-organism.

An important difference between the social organization of the termites and that of the other social insects is that all termite castes are bisexual. Unlike the ants, bees, and wasps, in which only the females serve as workers, with the termites there are male and female soldiers, workers, and reproductives.

The more primitive termites feed exclusively on wood and related plant products, which they digest with the aid of mutualistic protozoa and bacteria in their gut. The protozoan fauna of the gut is a remarkable assemblage of flagellates, representing several families and many genera. In the more specialized termites, the protozoan fauna is lost, but bacteria are present and perform the function of breaking down the celluloses. Many of these higher termites no longer eat wood as such, but use wood, leaves, and other plant products to serve as a base for raising fungi. The fungus, *Termitomyces* (a basidomycete) is exclusively associated with termites and does not occur outside their nests. The hyphae of the fungi grow on the termites' fecal pellets and convert the cellulose in the pellets to a form that is digestible by the termites.

Termites are closely related to cockroaches, and are often placed within the superorder Dictyoptera, which also includes the Blattaria and Mantodea. Like the cockroaches, the most primitive living termite, *Mastotermes darwiniensis*, has hind wings with small anal lobes and lays its eggs in an ootheca-like mass. More advanced termites do not possess anal lobes on their hind wings, nor do they produce such egg masses. Termites first appear in the fossil record in the Cretaceous.

Order Mantodea

The praying mantids are easily distinguished from most other insects by their raptorial forelegs (figs. 13.29). They characteristically rest upon the four posterior legs, which are adapted for walking and running, while holding the enlarged anterior legs together and extended forward. Some Hemiptera show a comparable adaptation of the forelegs, but are easily distinguished by their sucking mouthparts. Among the Neuroptera, the Mantispidae are extremely similar to the mantids, for which they may be mistaken, but the mantispids' characteristically membranous wings will easily separate the two orders.

The mantids may be briefly described as moderate-sized to large insects, with the North American species ranging from about 10 mm to over 100 mm in length. The anterior legs are strongly modified for grasping, with elongated coxae and the femora and tibiae usually heavily spined and capable of folding closely together to grasp prey. The tarsi are five-segmented. The head is freely movable and capable of considerable rotation. Mantids usually possess three ocelli, and the compound eyes are large. The mouthparts are situated beneath the head and rarely extended forward. The mandibles are strong and are adapted for dismembering prey. The antennae are usually short and filamentous. The prothorax is usually very long and freely movable, while the meso- and metathorax are roughly equal in length and more or less united. The wings are usually well developed, but sometimes are reduced or lacking. Forewings, if present, are tegminous, and overlap on the abdomen. The hind wings, if present, are membranous, with many veins radiating from the base, and fold fan-like beneath the tegmina. The abdomen is elongated, cylindrical, flattened, or oval, and the cerci are short and segmented. Eggs are laid inside a Styrofoam-like ootheca. Metamorphosis is paurometabolous, with the immature stages similar to adults in appearance and habits.

Most mantids are elongate and rather stick-like insects. Some tropical forms, however, have the thorax and abdomen expanded and the legs broadened and flattened, so that the resemblance to leaves is extraordinary. Other species have the wings marked and colored so that when extended, they simulate eyes, or the whole insect resembles a flower—a camouflage that enhances the mantids' predatory success.

Mantids have also evolved an ability to detect bats and respond with evasive behavior. There is an "ear" on the venter of the abdomen that can detect the bats' ultrasonic echolocation signals. When the predator's presence is revealed, the mantid will abruptly stop flying and drop to the ground, thereby avoiding the jaws of the hungry bat.

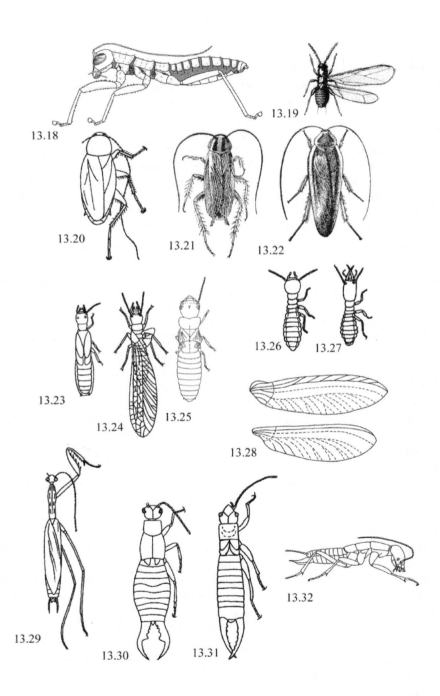

13.18 Heelwalker, Mantophasmatodea. (Drawing by Victoria Saxe)
13.19 Winged Zorapteran.
13.20 Cockroach, Blattaria.
13.21 American cockroach, Blattaria.
13.22 Wood roach, Blattaria.
13.23 Subreproductive termite, Isoptera.
13.24 Adult female termite, Isoptera.
13.25 Adult female termite with wing stubs, Isoptera.
13.26 Termite worker, Isoptera.
13.27 Soldier termite, Isoptera.
13.28 Termite fore and hind wings showing line of breakage.
13.29 Mantid adult, Mantodea.
13.30 Adult male earwig, Dermaptera.
13.31 Adult female earwig, Dermaptera.
13.32 Ice crawler adult female, Grylloblattodea.

Mantids appear to have evolved sexual cannibalism as part of their mating ritual, though it is not a requirement of mating. Males approach the larger females with great care and mate on top of her. This position makes the male more difficult for her to reach, but not impossible. Indeed, studies have shown that males may comprise up to 50% of the prey of females. If a male is decapitated by the female, the headless body will continue making mating movements, which may last up to 20 hours.

The mantids are closely related to the cockroaches with which they are sometimes united in the superorder Dictyoptera. The males of both mantids and roaches have similar asymmetrical genitalia, and the females of both produce ootheca. Fossil mantids have been found in the Cretaceous. These fossil mantids possess raptorial forelegs and a short pronotum. There is a fossil from the Jurassic that might be an early mantid, but more specimens and study are needed to be certain.

Order Dermaptera

The earwigs (figs. 13.30–13.31) are easily recognized by their characteristic forceps-like cerci and by the small, completely hardened elytra, which meet in a straight line down the back and resemble those of the Coleoptera.

The North American Dermaptera are small to moderate-sized insects, 4 mm to over 25 mm long, elongate and usually somewhat flattened in form. Their integument is heavily sclerotized and often deeply pigmented, brightly colored, or patterned. The surface is usually shining, but sometimes densely covered with hairs or small pits. The head is directed forward with chewing mouthparts that are blattoid in structure and long filiform antennae.

The compound eyes are often well developed, an[d the] are absent. The
simple prothorax is freely moveable, with the mes[o-] etathorax closely
united, and the metanotum is often fused to the fir[st abdomi-]nal tergum. The
legs are usually short, and the tarsi are three-segm[ented.] The wings, when
present, consist of hardened, truncated forewings (t[he elytra]) and semicircular
hind wings with a thickened basal area and radial ve[ins. The] membrane. The
hind wings, when not in use, fold completely bene[ath the el]ytra, folding first
fan-like and then transversely by two folds in r[ight-angle di]rections. When
folded beneath the elytra, a small tip of each hin[d wing is] exposed that is
textured and colored like the elytra. The ab[domen i]s terminated by
unsegmented cerci, which in adults usually take [the form] of pincers or
forceps. Metamorphosis is paurometabolous. [Earwigs] are omnivores,
feeding on both decaying and living plants as we[ll as othe]r insects. Most
earwigs are nocturnal, and apparently all are capab[le of limit]ed parental care,
brooding over the eggs and tending the first two ny[mphal sta]ges.

The earwigs use the forceps on their ab[domen f]or prey capture,
defense, and courtship displays. The forceps are u[sually mo]re curved in the
males (fig. 3.30) than in the females (fig. 3.31).

The name "earwig" is derived from the bel[ief that th]ese insects crawl
into human ears and cause insanity. In reality, mo[st species] of dermapterans
are inoffensive creatures. *Forficula auricularia* [in part] may sometimes
become a troublesome garden or house pest, but [due to its] habit of preying
upon other insects, it may actually be more usefu[l than d]amaging. Other
species may cause annoyance by entering houses.

Fossil Dermaptera date back to the late [Triassic a]nd early Jurassic
periods. These ancient earwigs have veins in their e[lytra, five]-segmented tarsi,
and long, segmented cerci that are not modified into [forceps.]

Order Grylloblattodea

The Grylloblattodea is a small order wit[h only 2]5 species in five
genera. They are known only from northwestern [United S]tates and eastern
Asia.

Grylloblattids (fig. 13.32) have small c[ompound eyes, no] ocelli, chewing
mouthparts, cursorial legs with five-segm[ented t]arsi, elongated
multisegmented cerci, and an unmodified, blade-li[ke ovipo]sitor. Wings are
lacking in all living members of the order. Th[ey live i]n areas of cool
temperatures, including caves and mountainous r[egions an]d feed on both
plant and animal matter. Development is paurome[tabolous]and the life cycle
may be very long, with the adult stage being reach[ed in a]s many as seven
years.

The grylloblattids have been linked to t[ermites,] earwigs, and
heelwalkers. Possible fossil grylloblattids from the P[ermian a]nd the Mesozoic
possessed four wings with an anal lobe on the hin[d wings.] Some authorities

believe that fossil grylloblattids, which now include several families of extinct insects originally placed in the extinct order Protorthoptera, may be the ancestors of the Polyneoptera.

CHAPTER 14: THE PARANEOPTERA

The subdivision Paraneoptera, sometimes referred to as the Hemipteroid assemblage, is a monophyletic group of orders (evolved from a common ancestor) that possess wings without complex venation, hind wings without an anal lobe, no cerci, no ocelli in the nymphal stages, and only a few Malpighian tubules. The subdivision also exhibits a trend towards cephalization (fusion of ganglia in the thorax) and the evolution of sucking mouthparts with a cibarium.

In the past, the Paraneoptera included six orders, but morphological and molecular studies have reduced that number to three. Such changes in taxonomy might seem trivial to the non-specialist, but the changes reflect our increasing understanding of the evolutionary relationships of the insects.

Order Psocodea

The Psocodea include the bark lice, book lice, biting lice, and sucking lice. These insects were previously placed in two orders, the Psocoptera (bark and book lice) and the Phthiraptera (chewing and sucking lice). Moreover, the latter order was sometimes split into two orders, the Mallophaga (chewing lice) and the Anoplura (sucking lice), but molecular analysis has indicated that the biting and sucking lice are part of a subgroup of the former Psocoptera; hence the change in the ordinal name to the Psocodea.

The bark lice (fig. 14.1) or book lice (the former Psocoptera) are small insects, usually less than 5 mm long. Most species are found outdoors, often on the bark of trees, as the name "bark lice" implies. They are usually feeders on fungi, cereals, pollen, insect fragments, or other dry materials. A few species enter buildings, where they are occasionally damaging to foodstuffs, books, and other material. Some species are semisocial, living in colonies that are sometimes covered by a silken web. Many genera are cosmopolitan or widespread.

The bodies of bark lice are short and relatively soft. Two pairs of wings are present in the adults of many species (fig. 14.2 and 14.3), but short-winged and apterous adults (book lice) are common (figs. 14.4 and 14.5). The head is large and freely articulated, with the mouthparts directed downwards. The compound eyes are large except in a few wingless forms, and there are usually three ocelli. The antennae are long, slender, and filiform or bristle-like, with up to 50 segments. The maxillary lacinia is modified to brace the head while feeding, and the mandibles are strong and toothed, with a grinding surface. The prothorax is usually small, whereas the meso- and metathorax are usually distinct, but sometimes fused. The wings are folded back over the body in repose and the forewings are larger than the hind wings, which are sometimes lacking. Wing venation (figs. 14.2 and 14.3) is reduced, with few or no cross-veins, but with one or more of the longitudinal veins usually strongly curved (fig. 14.2). They are sometimes very hairy or scaly. The abdomen is short and nine- or ten-segmented. Cerci are absent. The legs are adapted for walking, jumping, or running, and include a two- or three-segmented tarsus. Internally, bark and book lice have only four Malpighian tubules and no abdominal ganglia.

Some psyllids (Hemiptera) may be mistaken for bark lice, but the last segment of the antenna in the Psylloidea always has two characteristic setae.

The chewing lice (former Mallophaga) are small (< 5 mm), wingless ectoparasites found primarily on birds and mammals. Their mouthparts are adapted for biting and, except for one family, they seldom damage the skin of their hosts. Most species feed on hair, feathers, dead skin, or debris and are usually harmless unless present in large numbers.

Most chewing lice are flattened or elongate and cylindrical (fig. 14.6). The flattened forms generally live on the skin of the host beneath the feathers or hair, whereas the cylindrical forms live among the barbules of the large quills of birds, where they sometimes damage the feathers so badly that the birds cannot fly. This is a frequent problem in domestic pigeons. Compound eyes are small and ocelli are lacking. The mandibles are usually distinct and strong, with sharp edges. The rod-like maxillary lacinia are similar to those of the bark and book lice and are also used to brace the head while feeding. Cerci are lacking. The legs are usually more or less adapted for grasping hair or feathers, with the tarsi one- or two-segmented and bearing one or two claws. The elephant louse (fig. 14.7), *Haematomyzus elephantus* Piaget, an ectoparasite of both Asian and African elephants, possesses a proboscis with tiny mandibles at the tip.

Wild animals generally have only light infestations of biting lice unless they are diseased or injured. Under crowded conditions or when sick or injured, domestic mammals and birds may develop heavy infestations. The birds' habit of "dusting" themselves with ants (a process that often involves applying the ants' secretions of formic acid to the feathers) is important in suppressing biting lice populations.

The sucking lice (former Anoplura) are small (< 4 mm long), secondarily wingless, blood-feeding ectoparasites of mammals. They are common on rodents, but relatively rare on other herbivores and unknown from bats, marsupials, monotremes, and most carnivores, except the foxes, dogs, and pinnipedes.

All sucking lice, except the family Echinophthiriidae, have the tibia and one-segmented tarsus adapted for grasping hairs (figs. 14.8 and 14.9). The mouthparts are minute and adapted for piercing and sucking, with three slender stylets that are withdrawn into the head when not in use. The exact homologies with the mouthparts of other insects are questionable. The eyes are small, and the antennae are short and three-segmented in nymphs and some adults, but usually four- or five-segmented in the adult stage. The body is usually flattened, with the thoracic segments fused without evident sutures. Metamorphosis of all the lice is paurometabolous.

Humans are among the victims of the sucking lice. Pubic or crab lice, human body lice, and head lice are all annoying pests. Individuals who have access to frequent baths and clean clothes are not likely to be infested. Head lice have become frequent pests of our children and are easily spread by sharing combs, brushes, or hats. They are becoming resistant to the standard treatments, making control increasingly difficult. Most lice are merely irritating, but the human body louse is the vector of epidemic typhus.

Fossil representatives of primitive bark lice are found in Permian rocks. These fossils have wings of equal size, small cerci, and legs with four- or five-segmented tarsi. The hind wings had evolved to the typically smaller size by the Jurassic period. These Permian fossils are considered to be a stem group of Paraneoptera.

The fossil record of the chewing and sucking lice is poor. There is a putative early Cretaceous fossil louse that was 17 mm long, which had large compound eyes and legs with small claws. It appears to be an ectoparasite, but its host is unknown. The mammals of the early Cretaceous were quite small, and the general relationship in ectoparasite length with host size rules out these small mammals as this fossil's host. This large insect might have been an ectoparasite of pterosaurs, which had a hair-like covering, or of feathered dinosaurs.

Order Thysanoptera

The Thysanoptera, or thrips, represent a group of sucking insects distinct from the other hemipteroid orders in several basic characteristics. Most thrips are small insects about 0.5 to 5 mm long, but some tropical thrips attain a length of up to 14 mm. The winged adults are usually easily distinguished from other insects by the conical mouthparts, narrow wings fringed with long setae (figs. 14.10 and 14.11), and legs which usually end in a bladder-like or hoof-like enlargement, so that they appear to have no claws.

The thrips can be characterized as having an elongate and slender body that is somewhat flattened or cylindrical. The head is directed downward, with the mouthparts arising from the base. Distinct compound eyes are usually present and sometimes coarsely faceted. One or two pairs of ocelli are present in winged forms, but lacking in wingless forms. The clypeus is triangular and asymmetrical, and along with the labrum and labium forms a cone-like structure through which passes the elongated piercing left mandible and two maxillary stylets, which form a tube. Thrips puncture their food with the mandibular stylet, after which the maxillary stylets suck up fluids. The antennae are six- to nine-segmented and usually have conspicuous bristles and sense organs. The prothorax is usually free, with the meso- and metathorax fused together. The legs have one- or two-segmented tarsi equipped with one or two small claws, which are often inconspicuous because of an eversible bladder at the apex of the tarsus. This bladder can be expanded or retracted and assists the insect in adhering to slick surfaces. The forelegs are sometimes enlarged. The wings are narrow with few veins and a fringe of long setae, but are sometimes reduced or absent (figs. 14.13 and 14.14).

The thrips, as in the Hymenoptera, have a haplodiploid form of sex determination, in which females are produced from fertilized eggs and males from unfertilized eggs. This may have contributed to the evolution of truly social thrips that form galls on Australian acacias. (Galls are the result of abnormal plant growth caused by some kind of insect stimulus.) These thrips have male and female soldiers that will engage in altruistic behavior to protect their gall and their siblings.

Thrips are paurometabolous, but they exhibit some unique features in this development. The immature thrips (fig. 14.12) molt four or five times before the definitive adult molt. The first two stages are active nymphs that are generally similar to the adults except for the lack of wings. Towards the end of the second instar, some thrips spin a cocoon in which the insect molts to a resting nymphal stage called the "prepupa," which eventually molts to the "pupa." The "pupa" may not have antennae, or if it does, the antennae will have fewer segments than in the adult. The "pupa" also develops wing buds and undergoes some internal reorganization. This development shows similarities to the pupae of holometabolous insects, but thrips still are considered to be paurometabolous.

Most thrips are sap feeders, damaging crops and thereby becoming economic pests. A few species, however, feed on small insects such as aphids, and are beneficial.

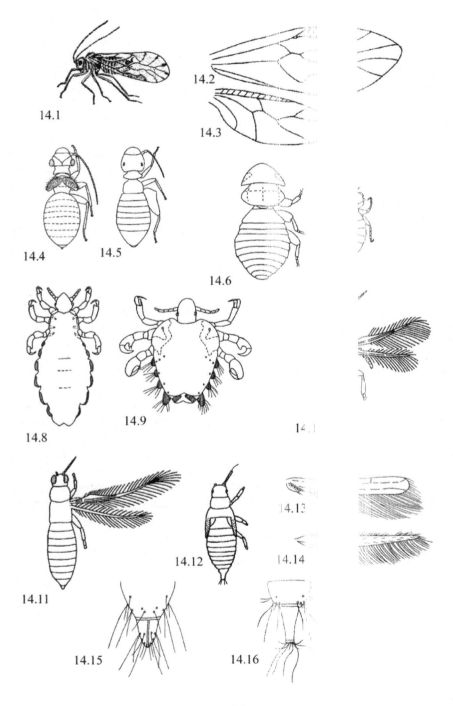

14.1 Adult winged bark louse, Psocodea.
14.2 Forewing of a bark louse, Psocodea. Note the strongly curved cubital loop in the forewing.
14.3 Hind wing of a bark louse, Psocodea.
14.4 Wingless bark louse, Psocodea.
14.5 Book louse adult, Psocodea.
14.6 Bird louse, Psocodea.
14.7 Elephant louse, *Haematomyzus elephantus*, Psocodea.
14.8 Human body louse, *Pediculus humanus*, Psocodea.
14.9 Crab louse, *Phthirus pubis*, Psocodea.
14.10 Adult thrips, Thysanoptera.
14.11 Adult thrips, Thysanoptera.
14.12 Nymphal thrips.
14.13 Forewing of thrips, *Aelothrips*.
14.14 Forewing of thrips, *Frankliniella*.
14.15 Tip of the abdomen of suborder Terebrantia.
14.16 Tip of the abdomen of suborder Tubulifera.

The thrips are divided into two suborders that are easily separated by the shape of the abdomen. The suborder Terebrantia (fig. 14.15) has a last abdominal segment that is short and blunt, whereas the suborder Tubulifera (fig. 14.16) has an elongated terminal abdominal segment that is shaped like a tube.

The thrips evolved from a now extinct group of Permian insects that were closely related to basal Psocodea. These insects had narrow wings with branching veins and no fringes. They had stylet-like mouthparts and legs without the terminal bladder. Thrips with fringed wings evolved by the Triassic, but these wings had venation similar to the Permian fossils, and more wing veins than modern thrips. The first thrips that can be assigned to modern families are found in the Cretaceous.

Order Hemiptera

The diverse Hemiptera possess mouthparts adapted to piercing and sucking. In the past, this order has been divided into two orders, with the Hemiptera comprising the true bugs and the Homoptera including the cicadas, leafhoppers, aphids, and scales. However, morphological and molecular studies have now shown that these insects belong in a single order with four suborders, the Heteroptera, Coleorrhyncha, Auchenorrhynca, and Sternorrhyncha.

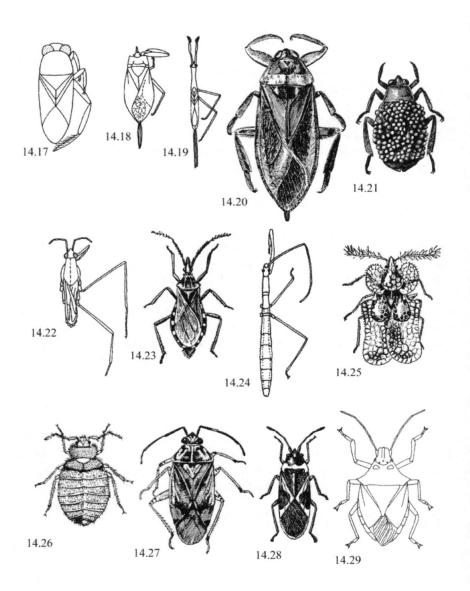

14.17 Backswimmer, Notonectidae.
14.18 Waterscorpion, Nepidae.
14.19 Waterscorpion, Nepidae.
14.20 Giant water bug, Belostomatidae.
14.21 A male giant water bug carrying eggs.
14.22 Water strider, Gerridae.
14.23 Assassin bug, Reduviidae.
14.24 Assassin bug, Reduviidae.
14.25 Lace bug, Tingidae.
14.26 Bed bug, Cimicidae.
14.27 Plant bug, Miridae.
14.28 Seed bug, Lygaeidae.
14.29 Stink bug, Pentatomidae.

The Heteroptera, or true bugs, have fully developed sucking mouthparts that articulate with the anterior part of the head and a well-developed cibarium. Winged heteropterans have forewings in the form of hemelytra, in which the base is leathery and opaque and the distal portion is membranous. Hind wings, when present, have greatly reduced venation. Tarsi are usually three-segmented, but rarely are reduced to one or two segments. Cerci are always lacking.

The Heteroptera include aquatic, amphibious, and terrestrial forms. The aquatic forms have their antennae carried in pockets on the head, and all but two families have raptorial forelegs that aid in the capture of prey. The Notonectidae, or backswimmers (fig. 14.17), swim upside down using their natatorial legs. The Nepidae, or waterscorpions (figs. 14.18 and 14.19), are easily identified by the long breathing tube at the tip of the abdomen. The Belostomatidae, or giant water bugs (fig. 14.20), are voracious predators of lakes and ponds where they feed on insects, snails, and sometimes small fish. The female lays her eggs on the back of the male for their protection (fig. 14.21).

The amphibious Heteroptera are predators of the water's surface and have evolved tarsi that enable the insects to move about on the surface tension of ponds and streams. The Gerridae, or water striders (fig. 14.22), can be often found scooting on the surface of quiet fresh water.

The terrestrial bugs include both predators and herbivorous forms. The Reduviidae, or assassin bugs (figs. 14.23 and 14.24), are predators of other insects, and can inflict a painful bite on humans. In South America, *Triatoma* is the vector of Chagas disease. The Tingidae, or lace bugs (fig. 14.25), are distinguished by the lace-like appearance of the surfaces of the head, wings, and pronotum. Lace bugs are plant feeders and some are horticultural pests. The Cimicidae, or bed bugs (fig. 14.26), are major urban pests. These secretive, wingless insects hide in tight spaces during the day and come out to feed at night. The Miridae, or plant bugs, is the largest family of Heteroptera, and includes many that are pests of vegetables and flowers. The Lygaeidae, or seed bugs (14.28), include many bugs that are black and orange or red. The bright colors warn potential predators that they are distasteful. The Pentatomidae, or stink bugs (fig. 14.29), have a characteristic shape and five-segmented antennae. Their name comes from the unpleasant odor they produce when they are under threat from predators or insect collectors.

The other suborders of the Hemiptera show the greatest diversity of body form and life history of any of the orders of insects considered so far. Their most characteristic common feature is the sucking mouthparts, which arise from the back of the head and usually project backward between the forelegs, sometimes seeming to arise from the thorax, in contrast to those of the Heteroptera, which arise from the front of the head (fig. 14.30). The wings, if present, are extremely variable in appearance. In some, such as the cicadas, the wings are membranous and often transparent, with strong veins. In others, such as the leafhoppers, the forewings may be uniformly leathery. Many lack wings entirely in one or the other sex. Male scale insects, for example, resemble delicate flies in having only the forewings well developed. Female and nymphal scale insects are particularly hard to characterize, since they may be wingless and bear little resemblance to other insects.

The mouthparts are usually distinctly elongated, with a rostrum formed by the clypeus, labrum, and labium enclosing the stylet-like mandibles and maxillae, which together form a sucking tube. In scale insects, however, only the stylets may be present, and these may be withdrawn into the body when not in use. All are plant feeders, but their habits are as diverse as their external appearances. They are found in grasslands, forests, and other habitats, even though many species are found in damp places, none are truly aquatic. Many are free-living, with nymphs and adults associated on the same plant, whereas others form galls or other protective coverings. Many have an alternation of sexual and asexual (parthenogenic) generations, which may be very different in appearance.

The suborder Coleorrhyncha is restricted to a few species that live in Australia and South America. Their southern distribution is considered evidence that they originated before the breakup of the ancient continent of Gondwanaland. The suborder Auchenorrhyncha includes the cicadas, the leafhoppers, and treehoppers. The Cicadidae, or cicadas (fig. 14.31), are common singers in the late summer, and the periodical cicadas emerge in considerable numbers in parts of the eastern United States once every 17 or 13 years. The Cicadellidae, or leafhoppers (fig. 14.32), are common plant feeders and include many horticultural pests. They can be easily identified by the row of spines on their hind tibiae. The Membracidae, or treehoppers (fig.

14.33), can be identified by their large pronotum, which takes many shapes and may mimic thorns. Some treehoppers are pests of apples and other trees.

The suborder Sternorrhyncha includes many insects of economic importance. The Aleyrodidae, or whiteflies (fig. 14.34), are minute insects that are sometimes found on the underside of leaves. Some whitefly species are pests of greenhouses and ornamental plants. The Aphidae, or aphids (fig. 14.35), can be easily identified by the two cornicles on the abdomen. They include many pests and have a complex life cycle. Many aphids overwinter as eggs, which hatch in the spring into wingless females. These females produce more wingless females parthenogenetically. In time, winged females are produced that can disperse to a new host plant, where they produce more wingless females. Later in the year, both winged males and females are produced; after mating, the females lay eggs to overwinter.

The Sternorrhyncan superfamily Coccoidea, or scale insects (figs. 14.36 and 14.37), is a large group of very small and specialized forms. Male scale insects look like small flies because the hind wings are reduced to structures that appear similar to the halteres of a true fly, but they can be differentiated by their lacking mouthparts and the style at the tip of the abdomen. Nymphal scale insects, called crawlers, have legs and antennae. After they molt to the second instar, they settle down and form a waxy scale that covers the body. The adult female does not develop wings, but remains under the scale, where she produces her eggs. The male develops his wings during a quiescent final instar. This stage of male growth is often called a pupa, even though scales have a paurametabolous development. The scales include many pests including the oystershell scale, the San Jose scale, and the European Elm scale. The lac scales are used to produce shellac and other varnishes.

The fossil history of the Hemiptera extends back to the Permian period. Many of the families evolved during the Mesozoic.

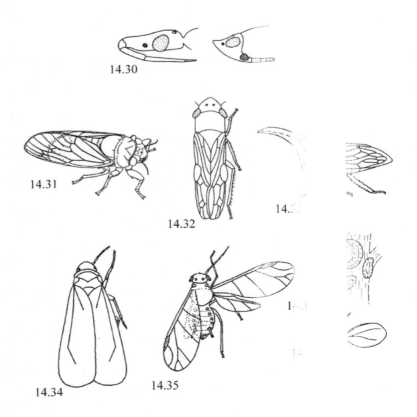

14.30 Comparison of heteropteran head with a ... rhynchan head. Note the point of articulation of the mouthpar... ie head.
14.31 Cicada, Cicadidae.
14.32 Leafhopper, Cicadellidae.
14.33 Treehopper, Membracidae.
14.34 Whitefly, Aleyrodidae.
14.35 Aphid adult with wings, Aphidae.
14.36 Scale insects on a tree, Coccoidea.
14.37 Male scale insect, Coccoidea.

CHAPTER 15: THE ENDOPTERYGOTA

The Endopterygota, sometimes called the Holometabola, are placed into a single subdivision because they have all evolved holometabolous development, which includes the egg, larval, pupal, and adult stages in the life cycle. They represent the most advanced living insects, though some of these orders retain many primitive features and have not surpassed the more highly specialized paurometabolous forms. In general, however, the endopterygote insects represent the last of the major trends in the evolution of insects, the origin of which can be traced back into Paleozoic times.

The origin of holometabolous development is not clear. However, hormonal work on development supports the view that larvae are extended stages between the embryo and the paurometabolous first instar. If this is the case, it seems likely that a polyneopteran may be ancestral to the Endopterygota. The most primitive holometabolous insects' larvae are somewhat similar in form to the adults, and adult forms possess a small anal lobe on the hind wing. They also retain an active pupal stage rather than an immobile one, suggesting a transitional form.

The Neuropterida

The Neuropterida include five living orders, three of which (the Megaloptera, Rhaphidioptera, and Neuroptera) are clearly closely related, and two orders (the Coleoptera and Strepsiptera) that form sister groups of the other three. There is some controversy regarding the placement of the Strepsiptera in this group, which will be discussed later.

Order Megaloptera

The Megaloptera, the dobsonflies and alderflies, are medium-sized to very large insects with four membranous wings and often pectinate antennae (fig. 15.1). The head is large and the mouthparts either project from the front of the head (prognathous) or are directed downward beneath the head (hypognathous). The mouthparts are adapted for chewing, with the mandibles sometimes greatly enlarged and saber-like in the males. Compound eyes are always present and ocelli are present in some. The antennae are elongate and many-segmented. Cerci are absent. The wings are leathery, but not tegminous, and the hind wings possess a small anal lobe, which folds fan-like when at rest. The costal cell has many cross-veins, but the branches of the veins that reach the margin seldom bifurcate (fig. 15.3). The costal and subcostal veins fuse before the tip of the wing. Metamorphosis is holometabolous, and the larvae are aquatic with tracheal gills on the abdominal segments that are developed from leg-like appendages. The pupae are terrestrial and free-moving.

The distinctive larvae of the Megaloptera are characteristic insects of streams and ponds. Larvae of the family Corydalidae are often used as fish bait and sold under the colloquial name "hellgrammites" (15.2). These larvae are predaceous, but the adults do not feed. Adult megalopterans differ from the Raphidioptera and Neuroptera primarily in the wing venation, but there are basic differences in body structures that indicate that these groups should be separated. Nearly all of the neuropterid insects differ from the other holometabolous orders in the extensive development of cross-veins in the costal cell.

There are two families in the order: the Corydalidae or dobsonflies, which are large insects with three ocelli, and the Sialidae, or alderflies, which are smaller and have no ocelli. Fossil megalopterans date back to the Late Permian with an extinct family that was likely a stem group of the living megalopterans.

Order Raphidioptera

The order Raphidioptera, or snakeflies, are moderate-sized predatory insects with an elongate pronotum and with four similar membranous wings (fig. 15.4). The head is large and nearly prognathous. Compound eyes are present and the ocelli may be present or absent. The mouthparts are adapted for chewing. The wings have many cross-veins and the marginal branches of the costal cell frequently bifurcate (fig. 15.5), and there is a conspicuous pterostigma beyond the costal cell. The costal and subcostal veins are not fused. The antennae are long and many-segmented. Cerci are absent, and the legs are adapted for walking. Metamorphosis is holometabolous, with the larvae being terrestrial and predaceous. The pupal stage is very primitive and bears some resemblance to a resting larval stage, but it is capable of movement.

The snakeflies' elongated pronotum distinguishes them from the Megaloptera and the Neuroptera, except for the neuropteran family Mantispidae. The Mantispidae have anterior legs modified for grasping much in the manner of the Mantodea, while in the Raphidioptera, the front legs are simply cursorial.

The mating behavior of the Raphidioptera is also primitive, with the male taking the inferior position during copulation. The larvae live beneath bark and in debris, where they prey upon other insects. Snakeflies occur on all the continents except Australia, but they are rarely collected.

The oldest Raphidiopteran fossils are found in early Jurassic rocks.

Order Neuroptera

The order Neuroptera includes many common, delicate insects, including the beneficial green lacewings. They are small to large insects with four membranous wings (fig. 15.6). The head is large and hypognathous, with well-developed biting mouthparts in the adult. The compound eyes are prominent and ocelli are sometimes present. The antennae are usually long, many-segmented, and threadlike, or sometimes clubbed. Cerci are absent. The wings usually have numerous cross-veins in the costal cell, and longitudinal veins are usually very numerous, frequently branching and conspicuously bifurcating along the wing margins (fig. 15.6). The fore and hind wings are usually similar, but the hind wings are sometimes greatly modified. Metamorphosis is holometabolous and the larvae are terrestrial except in one family, the Sisyridae, in which the larvae feed on freshwater sponges. The larval mandibles have a groove that connects them with the maxillae to form a sucking tube. There are eight Malpighian tubules, six of which are recurved backward and associated with the hindgut, where they function in silk production.

The antlions, lacewings, and other Neuroptera are extremely variable in comparison to the Raphidioptera and Megaloptera. In the Mantispidae, for example, the adults have raptorial forelegs for an active predatory life and closely resemble small mantids (fig. 15.7). Some Mantispids also resemble wasps. In a fine example of aggressive mimicry, they sit on flowers visited by other insects and apparently gain an advantage through their wasp-like appearance and behavior.

The larvae of the Neuroptera are also varied. The larvae of the Myrmeleontidae, or antlions (fig. 15.9), are adapted for living in sand or loose dry wood, in which some species construct funnel-shaped traps for other insects. The larvae of the Mantispidae are parasites in spider egg cases, which they seek out after hatching. Larvae of several families run about over vegetation, actively seeking scales, aphids, and other insects on which they prey. Neuropteran eggs are characteristically raised from the substrate on stalks, an adaptation to prevent early predation by siblings. The pupa is enclosed in a silken cocoon.

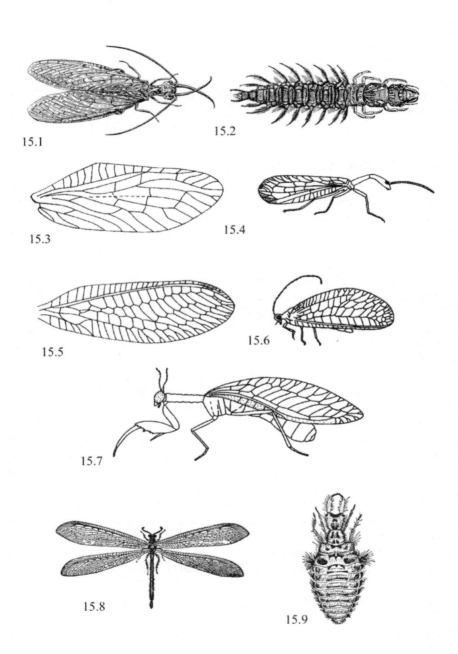

15.1 Male dobsonfly adult, Megaloptera (half its normal size).
15.2 Hellgramite, Megaloptera (half its normal size).
15.3 Megaloptera (Sialid) forewing.
15.4 Snakefly adult, Raphidiodea.
15.5 Chrysopid adult, Neuroptera.
15.6 Chrysopid adult at rest with wings folded tent-like over abdomen, Neuroptera.
15.7 Mantispid adult, Neuroptera.
15.8 Adult antlion, Neuroptera.
15.9 Larval antlion, Neuroptera.

The Neuroptera include several important families. The Chrysopidae (fig. 15.6) are the green lacewings, which are important natural predators that have been mass reared for release as a natural control of insect pests. The Myrmeleontidae, or antlions, have some larvae that construct conical pits to trap ants. The adults look superficially like dragonflies (fig. 15.8), but the antlions can fold their wings and the wing venation branches characteristically at the margins. The previously mentioned mantidflies (fig. 15.7), or Mantispidae, resemble tiny praying mantises with typical neuropteran wings.

Neuropteran fossils date to the Permian period, and suggest that the order was much more diverse in the past. Giant lacewings that appeared in the late Jurassic were the "butterflies" of the period. A mid-Cretaceous mantidfly evolved reduced forewing venation, with the hind wings modified into a haltere-like structure similar to that found in flies.

The Megaloptera, Rhaphidioptera, and the Neuroptera are often placed into a single order, the Neuroptera. The three are monophyletic, but differ in characters that are considered by many to warrant their placement into different orders within a single superorder. It is relatively easy to separate the three orders. The wing venation of the Megaloptera does not show terminal twigging; the Rhaphioptera have an elongate pronotum with cursorial forelegs; the Neuroptera generally do not have an elongated pronotum, and in the only exception (the mantispids), the forelegs are raptorial.

The Order Coleoptera

The Coleoptera, or beetles, are minute to very large insects with the forewings usually modified into hardened elytra that typically conjoin in a straight line down the dorsum, often completely shielding the hind wings and abdomen. The hind wings are occasionally absent, but when present, they are generally membranous, with a reduced but complex venation that allows the wing to fold upon itself in repose. The beetles usually have simple chewing mouthparts, with mandibles, maxillae, and a labium. The mouthparts are rarely reduced, but are sometimes situated at the end of an elongate rostrum.

The legs are adapted for running, walking, swimm... urrowing. Cerci are lacking. Metamorphosis is holometabolous (fi... and 15.11), with the larvae mandibulate and usually possessing well-... legs.

The beetles are the largest order of insec... ell over 350,000 species in the world and over 30,000 of these in ?... nerica. They are divided into four suborders. Two of the suborde... rchostemata and Myxophaga, include only a few species and a... collected. The Archostemata are the most primitive living beetle... tain remnants of their forewing's venation. They were much m... mon during the Jurassic than they are today and are considered to b... ossils.

The suborder Myxophaga contains 65 sp... ted to life in or around water, such as the edges of streams, rive... , and waterfalls. They are only 2 mm long and dorsoventrally fla... Their status as a separate suborder is being question, as they may l... ized members of the suborder Polyphaga.

The Adephaga are differentiated from th... uborders by the first abdominal sternite, which is divided by the la... (figs. 15.12 and 15.13). This suborder contains several familie... ing the familiar Carabidae (fig. 15.14), or ground beetles, some of... re typically black and are voracious predators. The Dytiscidae (f...), or predaceous diving beetles, are aquatic and carry their air with t... space under the elytra.

The suborder Polyphaga is the largest of t... uborders, and its members do not have the first abdominal sternite... by the coxae (fig. 15.11). This suborder includes many familiar famili... are best divided into five artificial groups for easier identification. T... roup includes the weevils, family Curculionidae (figs. 15.16 and... hich have their mouthparts situated at the end of prolonged snou... vast majority of weevils are herbivores, and some are economic pest...

The second group is the brachyelytrous l... hich have elytra that are shorter than the abdomen. This group i... he Staphylinidae, Histeridae, and Silphidae. The Staphylinidae, or ... ous rove beetles (figs. 15.18 and fig. 15.19), look superficially like ea... lack the forceps-like cerci. The Histeridae, or hister beetles (fig. 15.2... rrion feeders and have abruptly elbowed, clubbed antennae. The Sil... r burying beetles (figs. 15.21 and 15.22), are larger than hister beetles... softer elytra that are often characteristically marked with orange s... dult silphids are predators of insects found on carrion, whereas th... eed primarily on carrion but may also feed on plants and snails.

The beetles with clubbed antennae make u... d artificial group, which includes the Hydrophilidae, Coccinellidae, I... , Lucanidae, and Scarabaeidae. The Hydrophilidae, or water scav... tles (fig. 15.23), often have a long, keel-like spine that runs the leng... underside. Adult water scavenger beetles are mostly herbivorous, whereas the larvae are carnivores or cannibals. The Coccinellidae (figs. 15.24 and 15.25) are the ladybird beetles (sometimes called ladybugs) and have three tarsal segments on each leg. Most are predators, but there are some herbivorous exceptions. The Passalidae (fig. 15.26) are the so-called bessbugs, which are large black beetles that have a small L-shaped horn on the head. The passalids feed on rotting wood that contains bacteria. They also feed pre-chewed wood to their larvae. The Lucanidae, or stag beetles (fig. 15.27), are easily recognized by the males' large mandibles. Adult stag beetles feed on sap and the larvae bore into dead wood. The Scarabaeidae, or scarab beetles (figs. 15.28 -15.31), comprise a large family with many pests of gardens and turf as well as dung feeders. Common species include the Japanese beetles, June beetles, dung beetles, and the chafers.

The next group can be identified by the number of segments in the tarsi and includes the Meloidae and Tenebrionidae. The Meloidae are the blister beetles (fig. 15.32), so called because they produce cantharidin, a defensive secretion that can blister the skin. The meloids have a tarsal formula of 5-5-4, meaning that there are five tarsal segments on the front legs, five on the middle legs, and four on the hind legs. The head is typically wider than the pronotum. Meloids are herbivorous as adults, and their larvae are parasites of bees or eat grasshopper eggs. The Tenebrionidae, or darkling beetles (fig. 15.33), also have a 5-5-4 tarsal formula. They have a wide variety of body forms but can be easily distinguished from the meloids by their wider pronotum and notched eyes. Most darkling beetles are scavengers, but some are important pests of stored grains.

Two important families, the Cerambycidae and the Chrysomelidae, make up the next group. These families share a 4-4-4 tarsal formula and a bilobed third tarsal segment (fig. 15.34). The Cerambycidae, or long-horned beetles (fig. 15.35), have long antennae (at least half the length of the body) and emarginate eyes that look as if they wrap around the bases of the antennae. Some adult long-horned beetles feed harmlessly on pollen, but the larvae are considered pests of economic importance because they bore into the trunks of trees and feed on the woody tissues. The Chrysomelidae, or leaf beetles (figs. 15.36 and 15.37), have shorter antennae and their eyes are not emarginate. This family is herbivorous and includes many important pest species such as the rootworms, cucumber beetles, potato beetles, flea beetles, and tortoise beetles.

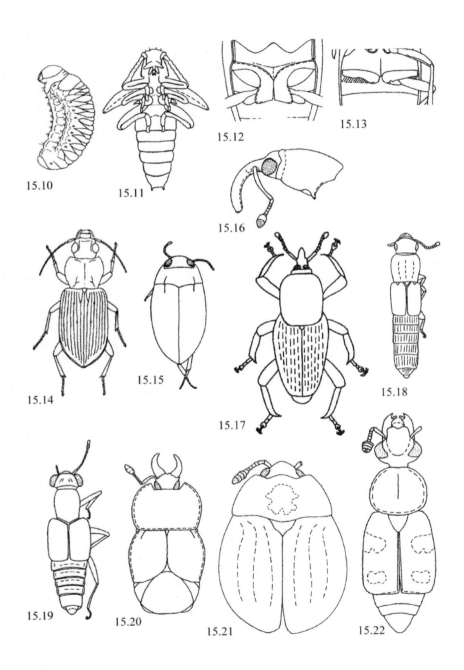

15.10 Coleopteran larva.
15.11 Coleopteran pupa.
15.12 The diagnostic first abdominal segment of an adephagan beetle.
15.13 The diagnostic first abdominal segment of an polyophagan beetle.
15.14 Carabidae, ground beetle.
15.15 Dytiscidae, predaceous diving beetle.
15.16 Head of a weevil (Curculionidae) showing antennal structure.
15.17 Curculionidae, weevil.
15.18 Staphylinidae, rove beetle.
15.19 Staphylinidae, rove beetle.
15.20 Histeridae, hister beetle.
15.21 Silphidae, carrion beetle.
15.22 Silphidae, carrion beetle.

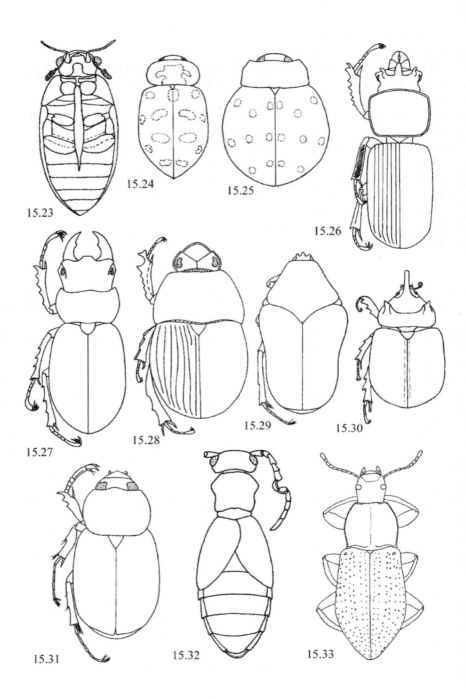

15.23 Hydrophilidae, water scavenger beetle.
15.24 Coccinellidae, ladybird beetle.
15.25 Coccinellidae, ladybird beetle.
15.26 Passalidae, "bessbug."
15.27 Lucanidae, stag beetle.
15.28 Scarabaeidiae, scarab beetle.
15.29 Scarabaeidae, scarab beetle.
15.30 Scarabaeidae, scarab beetle.
15.31 Scarabaeidae, scarab beetle.
15.32 Meloidae, blister beetle.
15.33 Tenebriondiae, darkling beetle, *Polopinus youngi*, named in honor of Frank N. Young, Jr.

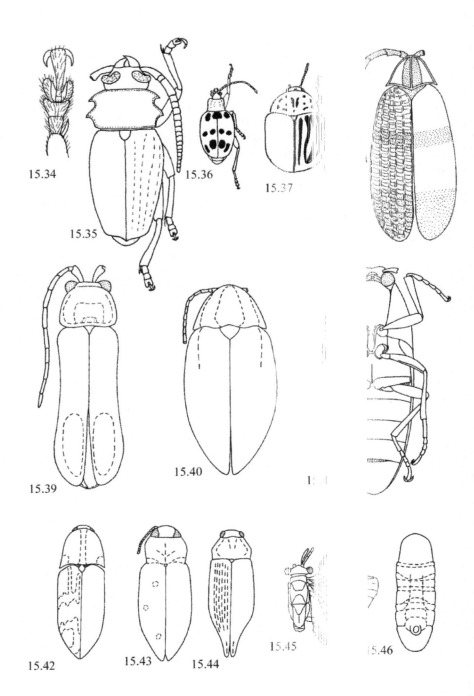

15.34 15.35 15.36 15.37

15.39 15.40

15.42 15.43 15.44 15.45 15.46

15.34 Bilobed tarsus of a cerambycid beetle.
15.35 Cerambycidae, long-horned beetle.
15.36 Chrysomelidae, leaf beetle.
15.37 Chrysomelidae, leaf beetle.
15.38 Lycidae, net-winged beetle.
15.39 Cantharidae, soldier beetle.
15.40 Lampyridae, firefly.
15.41 Elateridae, click beetle.
15.42 Underside of a click beetle, showing clicking mechanism.
15.43 Buprestidae, metallic wood-boring beetle.
15.44 Buprestidae, metallic wood-boring beetle.
15.43 Strepsiptera, adult male.
15.44 Strepsiptera, adult female.

The remaining common beetle families all have a 5-5-5 tarsal formula, but each family has some special characteristic. Three of these beetle families, the Lycidae, Cantharidae, and Lampyridae, have elytra that are soft and flexible. The Lycidae, or net-winged beetles (fig. 15.38), look like fireflies but have intricately sculptured elytra. Adult lycids feed on plant fluids or other insects, and the larvae are predaceous. The Cantharidae, or soldier beetles (fig. 15.39), have smooth, soft elytra and are commonly found on foliage or flowers, where they feed on pollen and nectar. Their larvae are mostly predators of insects, but a few herbivorous forms are known. The Lampyridae are the fireflies (fig. 15.40), which can produce light from specialized light organs. This light functions as a signaling device and is an important part of their mating behavior. The larvae are predaceous, feeding on insects, snails, and slugs.

Of the many remaining beetle families with a 5-5-5 tarsal formula, only the Elateridae and Buprestidae will be discussed. The Elateride (fig. 15.42), or click beetles, have typically hardened elytra and a uniquely shaped pronotum that partially conceals the head. Their name is derived from a clicking mechanism (fig. 15.43) formed by a prong on the prosternum that fits into a depression on the mesosternum. The parts of the mechanism fit snugly together so that when popped apart, an audible click is heard and the beetle may be propelled into the air, helping it to startle and escape from predators. Most adult click beetles do not feed, but their larvae (often called wireworms) live in soil and feed on roots. There are, however, some predaceous exceptions. Finally, the Buprestidae, or metallic wood-boring beetles (figs. 15.44 and 15.45), are usually metallic or brightly colored and often have highly sculptured elytra. Their larvae bore into wood, with the adults emerging to feed on foliage. This family includes many serious pests of orchards and woodlands including the Emerald Ash borer.

The beetles date back to the Permian period and may be one of the oldest holometabolous orders. Fossil beetles from the Permian have reticulated wings that still show the dense venation of the ancestral membranous condition. This suggests that the elytra of modern-day beetles evolved from a neuropteran-like ancestral wing, with the forewings becoming harder as the network of veins thickened and became fused.

Many beetle families are quite ancient. Carabid fossils are found in the Triassic, as are fossils of the rove beetles. Scarabs first appeared during the Jurassic and fossil dinosaur dung from the Cretaceous includes trace fossils of scarab burrows. The buprestid beetles and click beetles date back to the Late Jurassic, and the latter were quite diverse during that time. The net-winged beetles and soldier beetles are comparatively young, being found in amber from the Eocene, while the first definitive leaf beetles appear during the Cretaceous. The leaf beetles likely evolved simultaneously with the evolution of flowering plants, their major food source. Cretaceous ginger leaf fossils show the unique chewing marks produced by ginger-feeding leaf beetles. The oldest fossil weevils back to the Jurassic.

Order Strepsiptera

The strepsipterans are minute to small insects that are parasites of wasps, bees, thysanura, planthoppers, mantids, and roaches. An infested host can be identified by the protrusion of the puparium from between the abdominal segments. Adult male strepsipterans (fig. 15.45) have forewings that are reduced into strap-like pseudohalteres, while the hind wings are enlarged for flight. The eyes of the males are large, with the individual facets surrounded by a dense arrangement of hairs, and their antennae are lobed and covered with sense organs. Adult females (fig. 15.46), however, are wingless and remain within the abdomen of the host. Eggs are retained in the female, and the larvae are free-living as they search for a new host. After a new host is found, they molt and become legless second instars, and development continues within the host. When their metamorphosis is completed, the adult males emerge from the pupa, leave the host, and search for a host harboring a mature female with which to mate. The adult male lives only a few hours. The strepsipterans are rarely seen or collected. The best way to find them is to search for infested hosts.

The Strepsiptera is a controversial order of insects. In the past, it has been included within the Coleoptera as the family Stylopidae, placed as a sister group of the Diptera, or elevated to its own order. In this text, we have placed the order Strepsiptera close to the Coleoptera because they share many features with the beetles. For example, the wedge-shaped beetles (family Rhipiphoridae) have very short elytra and lobed antennae like the strepsipterans, and they are parasitic on many of the same host groups as the strepsipterans. However, the importance of these features in determining the strepsipteran relationships has been challenged. Unfortunately, molecular data is contradictory as to whether the strepsipterans are closer to beetles or flies. The true affinities of the order will have to await further research and more fossil material. The oldest fossil strepsipterans are larvae preserved in mid-Cretaceous amber.

CHAPTER 16: THE HYMENOPTERIDA

The superorder Hymenopterida contains only a single order, the Hymenoptera. This large order is of considerable importance with numerous pest species, such as the ants, and many beneficial species, including the honey bees. Knowledge of their diversity is growing, and there is evidence suggesting that eventually even more species of Hymenoptera than beetles may be described.

Order Hymenoptera

The bees, wasps, ants, sawflies, and other Hymenoptera are extremely diverse in body form. The heavily sclerotized body is characteristic of most members of this order, as is the marked constriction of the base of the abdomen that forms the narrow petiole, or "wasp waist." Two pairs of wings are present in most families, and the forewings are larger than the hind wings. The hind wings latch onto the posterior margin of the forewings with small hook-like structures called hamuli, forming a single lifting surface. The mouthparts are adapted for chewing, but typically also show modifications for lapping and sucking as well. The legs are usually all similar, with five-segmented tarsi. Females often have a distinct stinger or ovipositor on the abdomen.

There are two suborders, the Symphyta and Apocrita. The former can be identified by the broad attachment of the abdomen to the thorax (fig. 16.1), whereas the Aprocrita have a marked constriction between the abdomen and thorax (fig. 16.5).

The larvae are of two general types. In the suborder Symphyta, the larvae are similar to those of the Lepidoptera, with well-developed thoracic legs and often with fleshy abdominal prolegs. They can be distinguished from lepidopteran larvae by the lack of hooks or crochets on these prolegs. In the suborder Apocrita, however, the larvae lack legs, and although a head capsule is usually present, it may be very weakly sclerotized, especially in endoparasitic forms. Metamorphosis is holometabolous and the larvae may differ considerably at different stages. Endo-, ecto-, and social parasites occur in many groups, but truly aquatic larvae appear to be lacking. Some adults, however, may actively seek hosts for their larvae by "swimming" with their wings beneath the surface of water.

The pupae are generally similar to those of the Lepidoptera, with the head, antennae, legs, and developing wings usually apparent through the translucent cuticle. In some cases, the pupa is enclosed in a silken cocoon spun from the salivary silk glands of the larva.

The habits of Hymenoptera, both as adults and as larvae, are extremely varied. The Symphyta are primarily vegetation feeders through all stages of development. Many Apocrita, however, are external or internal parasites or parasitoids as larvae, larval gall-formers on plants, and predatory or nectar-feeding as adults. In the latter group, many predatory adults store food for the larvae (mass-provisioning) or even provide food as the larvae grow (progressive provisioning). In the Formicidae (ants), Vespidae (wasps), and some Apoidea (bees), a truly social organization comparable to that of the Isoptera is found.

The Formicidae are especially notable for their social behavior, which includes the growing of fungi for food, the tending of aphids and other insects for their honeydew or other exudates, and elaborate colonial behavior involving protection. The Dorylinae, or army ants, are an amazingly militant group found mainly in tropical forests, but some members of the genus *Eciton* extend into the temperate Nearctic.

The vast majority of insects in a hymenopteran colony are female and sterile. Moreover, these sterile females will protect the colony, even giving up their lives by attacking invaders. The evolution of this kind of social system is thought to be linked to the haplodiploid form of sex determination. Like the Thysanoptera, hymenopteran females are produced from fertilized eggs, whereas the males are produced from unfertilized eggs. The result of this form of sex determination is that sterile sisters will share more of their genes with their sisters than they would pass on by being fertile and reproducing. Therefore, by being sterile, caring for the queen's offspring, and even sacrificing their lives to protect the colony, the workers are, in fact, helping to ensure that more of their shared genes are being passed onto the next generation than if the workers could reproduce. Recently the view that these hymenopteran societies evolved through natural selection rather than the increased genetic relatedness resulting from haplodiploidy has been proposed.

Major families of the Hymenoptera include the sawflies (fig. 16.1), members of the suborder Symphyta and family Tenthredinidae, which are pests of conifers. The horntails (fig. 16.2), family Siricidae, are large symphytans that possess a "spear" of the tip of the [...] are capable of boring into the wood of both conifer[...] n. The horntails [...] iduous trees.

In the Apocrita, common families incl[...] Formicidae, Vespidae, Sphecidae, and Apidae. [...] ichneumon wasps (fig. 16.3), are slender, delicate [...] wing venation. The Formicidae (fig. 16.4) are the [...] by the pronounced petiole between the abdomen a[...] (fig. 16.5) include social wasps such as the hornets [...] fold their wings lengthwise. The Sphecidae (figs. [...] that cannot fold their wings lengthwise. The s[...] daubers and cicada killers, which are solitary and [...] insects or spiders. Finally, the Apidae, or bees (fig. [...] bees and honey bees. Honey bees were introduced [...] and they are of major economic importance. Their [...] basket, which is a hollowed-out portion of the leg [...] the packing of pollen grains (fig. 16.9). Ichneumonidae, ineumonidae, or with a distinctive are distinguished x. The Vespidae wjackets and can 16.7), are wasps nclude the mud their nests with clude the bumble country in 1621, possess a pollen bristles to aid in

The true relationship of the Hymenoptera [...] obscure. They are more closely related to the Pa[...] orders. Comparative morphology suggests tha[...] different from the rest of the holometabolous inse[...] Malpighian tubules and a unique wing venation. [...] fossils are Symphytans and date from the Triassic. [...] known from the Jurassic, while ants and bees evol[...] along with the flowering plants. t of the insects is han to the other lymenoptera are they have many st hymenopteran the Apocrita are g the Cretaceous

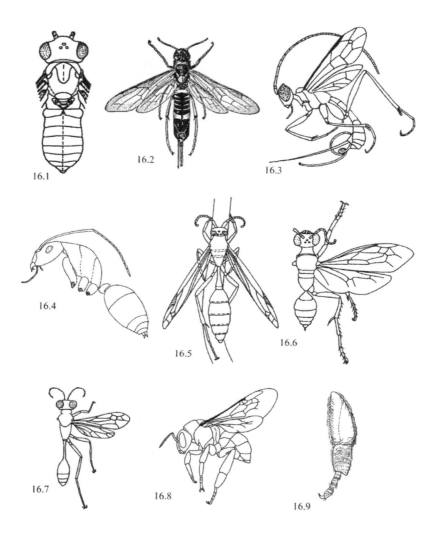

16.1 Tenthredinidae, sawfly showing broad attachment of abdomen to thorax.
16.2 Siricidae, horntail.
16.3 Ichneumonidae.
16.4 Formicidae, legs removed to show petiole.
16.5 Vespidae.
16.6 Sphecidae.
16.7 Sphecidae.
16.8 Apidae.
16.9 Hind leg of honey bee showing pollen basket.

CHAPTER 17: THE PANORPIDA

The superorder Panorpida includes five diverse orders, including the large orders Diptera (true flies) and Lepidoptera (butterflies and moths).

Order Mecoptera

The Mecoptera, commonly named scorpionflies or hangingflies, are small to medium-sized insects with four similar membranous wings (fig. 17.1 and 17.2) and chewing mouthparts situated at the end of an extension of the head called a rostrum (fig. 17.3). The antennae are long and slender, and the cerci are small with one or two segments. The compound eyes are well developed, and ocelli may be present or absent. Wings are usually present, with the radius extensively branched or rarely with many cross-veins and without a separate anal area. The prothorax is small, and the meso- and metathorax are similar in shape. The male genitalia are often elaborate, forming a large, reflexed bulb which resembles the stinger of a scorpion; hence the common name. Metamorphosis is holometabolus, and the larvae (fig. 17.4) of some scorpionflies resemble the caterpillars, except that they have evolved compound eyes with no more than 30 facets.

The scorpionflies are mostly found in dense vegetation. Some species are predaceous on other insects, whereas others are scavengers. They are secretive in their habits, often flying and hiding on the undersides of leaves when disturbed. The genus *Boreus* (fig. 17.5), or winter scorpionfly, is so named because it is often found running about on freshly fallen snow in winter.

The mating behavior of scorpionflies is complex, and there is considerable variation among the families. Males of many species present the female with the nuptial gift of a dead insect or a mass of saliva. This is followed by the release of a sex pheromone to draw the female to him. She will then accept or reject him based on the quality or size of the nuptial gift. Males of the genus *Bittacus* (Bittacidae), which hang suspended by their forelegs and grab their prey with their elongate hind legs, will sometimes steal nuptial gifts from other males. The *Boreus* male will capture the female and move about with her on his back while mating, using his modified wing pads to hold on to her.

The Mecoptera are known in the fossil record back to the Permian, when they were a much more diverse group than they are today. Mecopteran diversity flourished from the Permian through the Jurassic, but declined during the Cretaceous. Some of the fossil mecopterans were long-legged and quite bizarre, and they may have been ectoparasites of pterosaurs. Fossil material and molecular data support the view that the Mecoptera were the ancestors of flies and fleas.

Order Siphonaptera

The Siphonaptera, or fleas (fig. 17.6), are easily recognized by their laterally flattened, bristly bodies and by their enlarged hind legs, which are adapted for jumping. Adults of all species are secondarily wingless ectoparasites of birds or mammals.

Fleas may be briefly characterized as small insects, ranging from less than 1 mm to over 8 mm in length. The head is small, closely articulated to the thorax, and sometimes elongated. The antennae are short, thick, three-segmented, and recessed into grooves or depressions at the sides of the head when not in use. The compound eyes are poorly developed and often are greatly reduced or lacking. Ocelli are absent. The mouthparts are elongate and fitted for piercing and sucking, with the mandibles setiform, the maxilla blade-like, and the labial and maxillary palpi well developed. The thorax is small, while the abdomen is large, with one-segmented cerci. The legs are comparatively large and stout, with large coxae, and the tarsi are five-segmented and have two strong claws. Metamorphosis is holometabolous. The larvae (fig. 17.7) are cylindrical, legless, usually bristly, and free-living, with well-developed heads equipped with biting mouthparts. The pupa (fig. 17.8) is usually enclosed in a silken cocoon.

Fleas have played an important role in human history as the vector of the plague bacterium *Yersinia pestis*. In the 14th century, the plague of central Europe killed between 25% and 50% of the human population. The plague was spread by the rat fleas that left their dead rat hosts and then bit humans.

Recent DNA and morphological evidence places the fleas close to the mecopteran family Boridae, represented in North America by the winter scorpionfly. Both groups have multiple sex chromosomes and similar sperm morphology. Moreover, both boreids and fleas produce silken cocoons in which they pupate, and both can jump.

Primitive fleas are found in the Cretaceous, but they have long legs, and more segments in the antenna than do modern fleas. These fossils were of ectoparasites, but the host is unknown. Definitive fleas occur in the Miocene amber from the Dominican Republic.

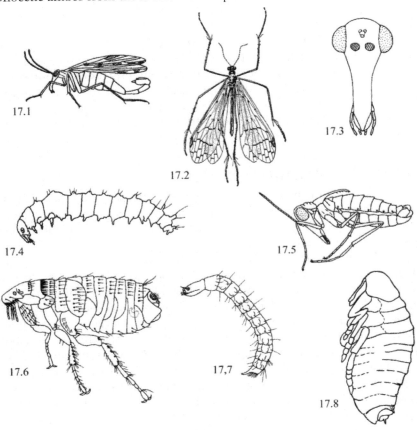

17.1 Mecoptera, scorpionfly: *Panorpa* male.
17.2 Hangingfly, Mecoptera.
17.3 Frontal view of scorpionfly head.
17.4 Mecopteran larva.
17.5 Mecoptera, scorpionfly: *Boreus* female.
17.6 Siphonapteran adult.
17.7 Siphonapteran larva.
17.8 Siphonapteran pupa

Order Diptera

The Diptera, or true flies, are distinguished from all other insects by the presence of only one pair of functional wings and by the knob-like balancing organs, or halteres, that evolved from the reduced hind wings. The mouthparts are modified for lapping or sucking, and chewing mandibles are never present, although the mouthparts are sometimes prolonged into a piercing proboscis. The labrum is often the principal piercing structure, but the mandibles and maxillae may also be stylet-like and perform the penetration. The legs are nearly always well developed, with five-segmented tarsi, and the flies are usually adept walkers and runners as well as fliers.

The larvae of the Diptera are variable in structure, but none possess true thoracic legs, and leg-like appendages, if present, are not jointed. The mouthparts are variously modified for biting, sucking, filter-feeding, or other functions. A sclerotized head capsule is present in many, but in some, only the hook-like jaws, which are retractable, allow them to be recognized as insects. Both aquatic and terrestrial forms may be equipped with groups of setae that are used for locomotion. Gills are present in some aquatic forms, although they may function as osmoregulators rather than organs of respiration. Other aquatic dipterans possess hemoglobin in the blood plasma, allowing for uptake of oxygen under nearly anaerobic conditions. Many aquatic larvae (fig. 17.9), however, respire by open spiracles that sometimes occur on long extensions of the abdomen.

The pupae of dipterans (fig. 17.10) are generally similar to those of the Lepidoptera. The developing eyes, mouthparts, and other appendages, closely appressed to the body, may be visible through the thin pupal skin. Strong bristles formed from large setae are often present. The pupae of the Chironomidae, Culicidae, and a few related families often have the thoracic region enlarged, with breathing trumpets on the dorsal surface and the abdomen slender, flexible, and capable of propelling the pupa actively through the water. In the higher Diptera, pupation takes place within the last larval skin (fig. 17.17), which becomes sclerotized and tanned so that the puparium (figs. 17.18 and 17.19) may more closely resemble a seed than an insect.

The higher classification of the Diptera is in a state of flux. Generally there are two recognized suborders, the Nematocera and Brachycera. The Nematocera are generally small and delicate flies that have antennae with at least six segments and are usually longer than the thorax. This suborder includes the Culicidae, or mosquitoes (fig. 17.11), the Chironomids, or midges (fig. 17.13), and the Tipulidae, or crane flies (fig. 17.12). Chironomids and crane flies superficially resemble mosquitoes to the extent that crane flies are often wrongly called "giant mosquitoes." However, both families lack the biting and sucking mouthparts found in the Culicidae.

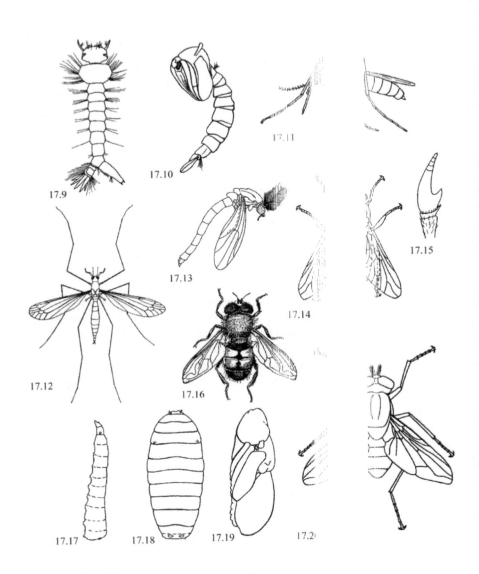

17.9 Mosquito larva, Diptera, Culicidae.
17.10 Mosquito pupa, Diptera, Culicidae.
17.11 Mosquito adult, Diptera, Culicidae.
17.12 Crane fly, Diptera, Tipulidae.
17.13 Lateral view of a midge, Diptera, Chironomidae.
17.14 Horse fly, Diptera, Tabanidae.
17.15 Horse fly antenna, Diptera, Tabanidae.
17.16 Syrphid fly, Diptera, Syrphidae.
17.17 House fly maggot, Diptera, Muscidae.
17.18 House fly puparium, Diptera, Muscidae.
17.19 House fly pupa, Diptera, Muscidae.
17.20 House fly adult, Diptera, Muscidae.

Flies of the suborder Brachycera are robust and have short antennae with fewer than six free segments (fig. 17.15). The suborder includes the horse flies and deer flies (fig. 17.14), family Tabanidae, which can be easily distinguished from other flies by their unique antennae and the V-shaped veins at the tip of the wings.

The Cyclorrhapha, a group within the Brachycera, have maggot-like larvae and their pupa develop within a puparium. This suborder includes the Syrphidae, Muscidae, Calliphoridae, and Tachinidae. The Syrphidae (fig. 17.16) are mimics of bees and wasps; the Muscidae are the familiar house flies (fig. 17.20); the Calliphoridae, or blow flies, are characteristically metallic and bottle-green; and the Tachinidae are large and have grey and black thoracic stripes.

The Diptera include many insects that are of medical and veterinary importance. Mosquitoes, for example, are vectors of several diseases, including yellow fever, encephalitis, and malaria. Malaria is caused by four species of the protozoan genus *Plasmodium*, and is spread by *Anopheles* mosquitoes. It is the cause of over two million human deaths each year. There is active research in the control of the *Anopheles* mosquito in hopes of limiting the spread of malaria. This may be achieved by developing methods of genetically modifying the *Anopheles* mosquito so that it cannot harbor *Plasmodium*.

The flies likely evolved from Permian mecopterans. Some of these fossil mecopterans possessed hind wings that were half the length of the forewing. The Brachycera appeared in the Late Triassic and diversified during the subsequent Jurassic period. The Cyclorrhapha are present in the early Cretaceous, but the major evolution of the higher flies occurred in the Tertiary.

Order Trichoptera

The Trichoptera (fig. 17.21), or caddisflies, are usually rather fragile insects with four membranous wings (fig. 17.22) and long, threadlike antennae. Some species closely resemble moths, except that instead of the characteristic scales of the Lepidoptera, the wings are covered by abundant hair-like setae. The forewing is usually rather narrow, but the hind wing may be broad, with a distinct anal area. In many genera, however, both wings are narrow, and in one family, the hind wings are fringed with long setae.

The larvae (fig. 17.23) of Trichoptera, with a few minor exceptions, are aquatic and often are case-builders (figs. 17.24-17.26), constructing protective coverings of pebbles and twigs or pine needles held together with salivary secretions. They are somewhat caterpillar-like, but have only a single pair of prolegs, with recurved hooks on the last segment of the abdomen.

The pupae are immediately distinguishable from those of the Lepidoptera by the large, functional mandibles with which the adults chew their way out of the pupal case. Also, unlike many Lepidoptera, the wing cases, legs, and antennae are more or less free. The pupa always occurs in a case: either the modified larval case, or a new case spun by the mature larva. The case is usually underwater and circulation of water though it seems to be necessary for survival of the pupa.

The oldest trichopteran fossils date back to the Early Jurassic period and were found in Germany. Tubular trichopteran cases from the mid-Jurassic constructed of silk, sand grains, and plant fragments have been found in Mongolia.

Order Lepidoptera

The butterflies and moths are usually among the easiest of insects to recognize due to the flattened, striate scales (modified from setae) on the wings and body. Even in a few species in which the wing membrane is transparent, flattened scales occur on some parts of the wings or body. The males of all Nearctic species are winged and capable of flight, but a few species have wingless or short-winged females. In some species, the scales are colored and may form bright and distinctive patterns.

The Lepidoptera are closely related to the Trichoptera, but in addition to flattened scales, the adult butterfly or moth usually possesses a coiled proboscis beneath the head. The adult trichopteran or caddisfly generally has an abundance of hair-like setae on the wings and body, and a coiled proboscis is never present.

The adult lepidopteran head is usually large, with compound eyes that sometimes have setae on or between the facets or long setae along the edge and overlapping the eye. Ocelli are often present and are located just above the compound eyes, behind the antennae. The antennae are long and many-segmented and variously modified. Mouthparts usually include a coiled proboscis composed of the conjoined galea of the maxillae, but they may be greatly reduced or lacking in some groups. The labial palpi are usually well developed and three-segmented. The maxillary palpi are often reduced or lacking, but are sometimes well developed and five-segemented. Functional mandibles are present only in Micropterygidae, a basal lepidopteran family.

The larvae of Lepidoptera are caterpillars with strong biting mouthparts, well-developed thoracic legs, and stumpy leg-like structures called prolegs on several abdominal segments (figs. 17.27 and 17.28). The larvae can be differentiated from those of other orders by the presence of tiny hooks (or crochets) on the prolegs. Nearly all are terrestrial herbivores, but some are aquatic, and a few are predators or parasites of other insects. Some leaf-mining and boring forms have the locomotory appendages reduced or lacking. None of the aquatic Lepidoptera build cases, but a number of terrestrial forms do, including the bagworm (fig. 17.29), which is a common pest of ornamental trees.

The pupae (fig. 17.30) of Lepidoptera are also similar to those of the Trichoptera, but mandibles are lacking in all except the family Micropterygidae. The lepidopteran pupa is usually formed in an underground chamber, among debris, or is covered with a cocoon spun by the larva. In some butterflies the pupa is naked, and suspended openly on vegetation.

The wings (fig. 17.31) are usually well developed, with many longitudinal veins but few cross-veins, except in some small species with narrow or linear wings in which the veins are greatly reduced. Fore and hind wings are usually somewhat different in size, venation, or both; the hind wing is usually smaller. Wings in small forms often have a distinctive fringe of setae that serves as an airfoil. Most moths have a strong spine or spines called the frenulum at the base of the hind wing, which engages with specialized spines on the forewing to hold the wings together in flight. A few moths have a finger-like projection, the jugum, on the inner margin of the forewing, which is overlapped by the hind wing to accomplish the same function as the frenulum. Butterflies and some moths have neither a jugum nor a frenulum, but the wings broadly overlap in a condition called amplexiform coupling.

There are several important families in the Lepidoptera. The butterflies comprise only about 6% of the species of the order, but their diurnal habits and bright colors make them quite noticeable. They are easily differentiated from moths by the clubbed tips of the antennae. Moreover, many butterflies produce a brightly colored pupa, or chrysalis. Some common families include the Nymphalidae, Papilionidae, and Pieridae. The Nymphalidae, the brush-footed butterflies, include the orange-and-black monarch (fig. 17.32), the red admiral (17.33), and the great spangled fritillary (17.34). The Papilionidae, or swallowtails (fig. 17.35), have a fingerlike tail on the hind wings. The Pieridae (fig. 17.36) are commonly referred to as the sulfurs, and many are bright yellow or white butterflies.

The family Hesperiidae, or skippers, (fig. 17.37) have antennae that end in small hooks. They combine features of both the moths and the butterflies and appear to be transitional between the two groups.

Moths can be separated from butterflies by their antennae (which are anything but clubbed), by forming a cocoon rather than a chrysalis, and by their generally nocturnal habits. Some major moth families include the Saturniidae, Noctuidae, Sphingidae, and Psychidae. The Saturniidae, or silkworm moths (fig. 17.38), are large, colorful moths with plumed antennae. The Noctuidae, or millers (fig. 17.39), are a large family of usually dark or mottled moths that sometimes have colorful hind wings, which are hidden by the forewings when at rest. The Sphingidae, or sphinx moths and hawk moths (fig. 17.40), have characteristic narrow, delta-shaped wings. The larvae of Psychidae, or bagworms (fig. 17.29), construct the bags in which they develop from silk and plant matter. Adult bagworm females remain in the bag, whereas the male develops wings and leaves his bag to search for the female.

The Lepidoptera likely shared a common ancestor with the Trichoptera. Both orders have an unusual method of sex determination in that males are produced from the XX chromosomes and females from XY chromosomes. The oldest Lepidoptera fossils are wings from the Jurassic. These fossil wings were covered with scales and had a wing venation that is shared with the Trichoptera. The Lepidoptera increased in diversity in the Cretaceous, apparently coevolving with the flowering plants.

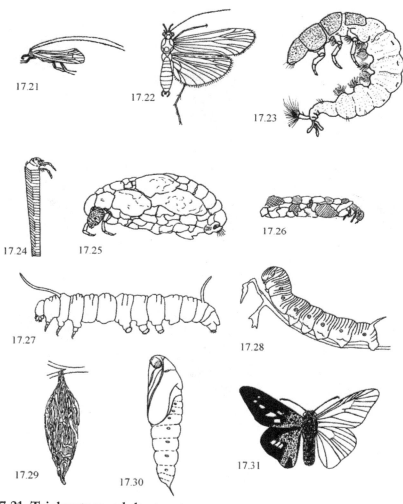

17.21 Trichoptera, adult at rest.
17.22 Trichoptera, adult with wings spread.
17.23 Caddisfly larva.
17.24 Caddisfly larva in its case of plant matter.
17.25 Caddisfly larva in its case of small pebbles.
17.26 Caddisfly larva in its case of small pebbles.
17.27 Lepidopteran larva, Sphingidae.
17.28 Lepidopteran larva, Nymphalidae.
17.29 Lepidoptera, Psychidae: bagworm.
17.30 Lepidopteran pupa, Noctuidae.
17.31 Lepidopteran adult with pattern-forming scales on left wing and scales on right wing removed, showing wing venation.

17.32 Nymphalidae, Monarch.
17.33 Nymphalidae, Red admiral.
17.34 Nymphalidae, Great spangled fritillary.
17.35 Papilionidae, Tiger swallowtail.
17.36 Pieridae, Sulfur.
17.37 Hesperiidae, Silver-spotted skipper.
17.38 Saturniidae, Polyphemus moth.
17.39 Noctuidae, Widow underwing.
17.40 Sphingidae, Tomato hornworm moth.

APPENDIX 1: METHODS OF COLLECTING AND PRESERVING ARTHROPODS

An old French recipe for rabbit stew begins with the exhortation, "First catch your hare." Obviously, before we can study insects, we must have insects to study, and preferably a series of specimens rather than a single individual of each kind. Although catching them may be a challenge, finding insects and other arthropods is very easy. In fact, some insects, such as the mosquitoes, will actually find you. It is only when seeking particular insects or making a representative collection that we must use specialized collecting techniques. Only a few methods will be outlined here. Most are suitable only during warm weather, but others can be used at any time of the year. Some methods can be quantified for ecological and special studies.

Picking: Many insects and other arthropods can be found on the ground or vegetation by simply looking for them and picking them up with the fingers or a pair of forceps. Turning over stones, logs, trash piles, and other accumulations often produces many arthropods, which can be easily caught. This is the best method during cold or rainy periods. Look carefully at tree trunks and on the undersides of rocks and logs, because many insects that are found in such places are camouflaged to resemble their background.

Beating: One of the most productive collecting methods is called beating and can be performed with either a net or a square piece of cloth or plastic. In beating with a net, the net is swung upward under trees, bushes, and other plants so that the ring and handle strike some moderately large branches and jar any insects resting among the leaves or on the twigs, causing them to fall into the net. If there is no net handy, one can construct an "umbrella" with two square feet of cloth or plastic, four safety pins, and two sticks about 2.6 feet long. Cross the sticks to form an "X" and safety-pin them to the corners of the cloth or plastic, leaving the latter hanging down in a shallow bag beneath. Grasp the sticks where they cross and hold the "umbrella" under a bush or other plant and beat the foliage and limbs with another stick. Both of these methods are surprisingly productive, particularly in the morning after a cool night following a warm day. Some entomologists actually use an umbrella, sometimes one that is mounted onto the handle at right angle.

Sweeping: Probably the most satisfactory general method of collecting is sweeping, except in bad weather. This requires an insect net, which is simply a cloth bag mounted on a metal ring attached to a handle. The net is swung repeatedly through grass or other vegetation and the contents examined. The rate and power of the strokes will determine how much extraneous vegetation is picked up. It is usually best to put the entire contents of the net into a killing jar or dump it into alcohol, because simply looking down into the open net lets most swift-flying insects escape and smaller forms are easily overlooked. Sweeping over flowers, such as asters and goldenrod, will often produce large numbers of a wide variety of insects.

Netting: Butterflies, moths, wasps, many flies, dragonflies, grasshoppers, mayflies, and many other insects must usually be netted individually or in small groups. Some fly too fast to be caught by hand or by sweeping or beating; others are ruined by being rubbed by vegetation or other insects. It is sometimes best to stand in one place when netting insects. Choose a place by some flowering plant or tree and let the insects come to you rather than running after them.

Aquatic Netting: Sweeping a wire sieve or a special aquatic net through vegetation along the edges of ponds, streams, or other bodies of water is an effective way of collecting many kinds of aquatic insects. Few insects, except some large forms, will be found out in open water. Most will be near the shore among vegetation, debris, or even in sand or silt at the margin. In bogs and springs, pressing the strainer down into moss or peat and letting the water rush in often produces interesting insects. Along the shores of lakes, ponds, and streams, throwing water on the damp marginal sand or silt causes many insects to "pop" out of burrows and onto the surface, where they can be picked up or pinned down with the strainer.

Separating: Many insects live in decaying leaf mold, dead wood, soil, trash, and other accumulations. Several methods are effective in separating these insects from their habitat. An easy method is to rake dead leaves or other debris into a pile, leaving a clear area around the edges. Pour some household ammonia on the center and place a plastic sheet over the pile. Beetles and other insects will often quickly appear in the cleared area, moving away from the gas.

Another common method of separating is to collect in a plastic bag a sample of debris, dead wood, decaying mushrooms, lichens, moss, trash, soil, or other material in which you suspect there may be insects. Later, place the material on the screen of a Berlese funnel. This is a large funnel with a coarse screen in the top and provision for a killing bottle or alcohol container under the smaller end. Many arthropods react quickly to changes in humidity, and as the sample dries, they will move down, fall through the screen, and end up in the container beneath. Live insects for study can also be collected in this way.

Finally, many small arthropods in debris or soil can be collected by sieving or screening the material over a piece of cloth or plastic. The sieve is filled with the material to be separated and lightly shaken. The insects that fall through are picked up before they can retreat back into the material.

Trapping: Many kinds of traps are used in collecting insects. Pitfall traps, which consist of cans, cups, or jars buried in the ground so that the opening is flush with the surface, are very effective. They may be baited with diluted molasses, honey water, dung, dead flesh, decaying mushrooms, and other attractants, but empty cans or jars may produce good results. Light traps are also effective. Those with ultraviolet bulbs or colored bulbs attract different insects from those with white bulbs. Dry ice is an effective attractant for mosquitoes and can be used for sampling where light is not permissible.

Attractants: Many substances can be used to lure insects within reach of the collector. Molasses, water, and a little rum or stale beer smeared on a tree trunk will attract moths at night during the summer months. With a flashlight or headlamp, the collector can usually catch them at this bait by simply slipping a killing jar over them. Light is also an attractant, and many kinds of insects can be collected around filling stations, shopping malls, neon signs, or other lighted places at night. Even in the daytime, an isolated filling station that is lighted at night may have moths, beetles, flies, and many other insects that were attracted to light the previous night resting on walls or beneath debris along curbs or other impediments.

Killing, Preserving, and Preparing Insects for Study

Killing and Preserving: Nearly all arthropods, including butterflies, moths, and flies, can be killed and preserved by putting them into a jar partly filled with 70% isopropyl alcohol or ethyl alcohol. If removed within a few days, all will remain identifiable, but after a longer time butterflies, moths, and flies become discolored, and it is difficult to mount them for study. In consequence, a dry killing jar is advantageous. One can be easily made by pouring plaster of Paris into the bottom of a glass jar, allowing it to dry, and then saturating it with ethyl acetate (which is often sold as "acetone-free" nail polish remover). Freezing insects is another easy method of killing them. Simply collect the insects in a container and place it in a freezer overnight. Insects can be stored for long periods of time in the freezer without drying out and becoming brittle.

Preparation for Study: Most arthropods can be studied effectively as fresh specimens or simply preserved in alcohol (70-85%). With a long series of specimens that must each be examined in a fixed position, however, liquid preservation may be inconvenient. Most insects, except the apterygotes, scale insects, other soft-bodied forms, and minute forms, are best mounted dry on pins or points. This is not done, however, for any arthropods other than insects. A pinned spider, mite, scorpion, crayfish, or other non-insect is usually unidentifiable since the structures that must be examined are lost or distorted in drying.

Pinning is fairly easily accomplished with most large insects when they are freshly killed or after they have been relaxed in a closed jar with a dampened paper towel or cotton. The pin is inserted from the dorsal surface downward through certain specific parts of the body, depending upon the order. Check Fig. A.1 for the proper position of the pin. Never, for example, pin a beetle through the thorax. The pin should be inserted through the right elytron near the base. With most other orders, however, the pin should go through the thorax, slightly to the right of the mid-dorsal line.

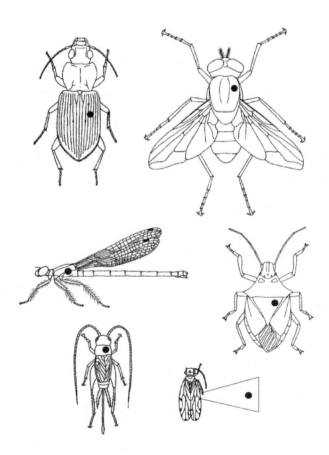

Fig. A.1 Pin placement for selected orders. The bottom right shows the orientation of the point for small insects.

Pointing (Fig. A.1) is simply the process of gluing an insect on a small, elongate triangle of heavy paper so that the specimen will not be damaged or destroyed by the pin itself. Various adhesives are acceptable, but shellac, Elmer's glue, or colorless fingernail polish are among the best. Use the smallest point and the least amount of adhesive that will support the insect. Glue the specimen in place so that a minimum of its surface is concealed. A specimen slapped on a point with a large blob of glue is worthless until removed, cleaned and remounted. The position in which a specimen is placed on a point is important. The head should always be pointed forward, with the point glued to the insect's right underside and extended to the left side of the pin. Right-handed or backward specimens increase the difficulty of comparing series. Some liberty can be taken with the position of the rest of the body.

Spreading is the process of extending the wings of an insect so that they may be examined more easily. This is usually done with all butterflies and moths and can be done with many other insects such as grasshoppers, wasps, large flies, and others. Many moths and butterflies cannot be identified after they have dried unless the wings are spread to expose the color pattern and wing venation. Spreading boards are easily made by cutting a long V-shaped groove in a thick piece of Styrofoam, balsa wood, or other soft material. The pin must be deeply embedded so that the insect will be at the desired height, because after the body dries, the pin cannot be moved without damaging the specimen.

Setting, or the arrangement of the legs in an extended position, is not generally done except for aesthetic reasons. However, arrangement of legs and antennae closely against the body of an insect is a protective measure, which should be carried out to prevent their breakage or loss. Mantids, walking sticks, and other long-bodied insects are best dried in tubes made by rolling soft paper around a pencil or larger cylinder, if necessary, and tucking in the ends. The insect is then carefully dropped into the paper cylinder head-first and the antennae and legs will be "trained" against the body. Great care must be taken not to damage such specimens in pinning if they are allowed to dry completely.

Slide mounts may be necessary for the study of some insects. Fleas, lice, scale insects, and others need to be cleared (have pigment removed) so that they may be viewed with the compound microscope at high magnifications. Springtails and many smaller insects and their larvae may be mounted directly in balsam or synthetic media, but all must be dehydrated by running them through 70, 80, 95, and 100% alcohol and then passing them into xylene before applying the mounting medium. Allow at least five minutes in each stage of dehydration or the insect may collapse and be ruined for study.

Labeling is an essential part of preparing a ... ...or study. If you follow the expedient of preparing a card or ... entry for each collection and numbering them serially, the number... used to identify the specimens until labels can be prepared. With ... ...nt ease of word processing, specific labels for large numbers of sp... can be prepared cheaply. Simply type out the information in a w... ...essor, copy, and paste in as many copies as needed. Printing shoul... ...pproximately six- point font to make the labels small enough for pin... alternative is to prepare partially blank labels, which give all necessa... ...ation except the date and catalogue number, which may be written... ...a fine pen. If insects are identified to species, a determination la... the name of the identifier should be added. See examples of labels b...

OH Hamilton Co.	IN Monroe Co.	(	lton Co
Cincinnati – along the	Bloomington	(	
Ohio River IX.1.2011	IX. 1.2011		
G Kritsky, coll.	F N Young, coll.	(	, coll.

Labels should be placed below the speci... the pins and so situated that they give the maximum protection to ... ...cimen when it is handled. With pointed specimens, the label shoul... to the left of the pin, beneath the point, and read from the upper left... ...inned specimens, the label should be placed beneath the specimen, r... ...m either the rear left or front right, but in any case, consistently.

Dried pinned insects are very fragile, and ... ...ould be taken to hold only the head of the pin or its tip, and to ... ...a soft surface to reinsert the pin. A kneaded gum eraser is handy for ... ing insects under a stereo dissecting microscope. Either the tip or t... ...f the pin may be inserted in the soft material at any angle so the inse... ...e examined from all points of view.

Insect study specimens are usually stored i... boxes or trays in glass-topped cabinet drawers. After drying, ins... very quickly if exposed to light, so the boxes or cabinets sho... light-tight. For temporary storage, cigar boxes (now available ... ...ially as "school boxes") are convenient. They should be equipped ... ...oft cardboard or foam bottom and, if stored for any length of ti... ...mall quantity of naphthalene flakes or paradicholorbenzene (moth ... ...lakes) should be put into each box every three months. In warm... ...egions, a mildew inhibitor should be sprayed into boxes or cabinet... before they are used, unless air conditioning is constant.

Arranging a collection is, for the most par... ...rerogative of the collector. However, a neat arrangement adds ... usefulness of a collection. This involves not only placing the speci... ...eat rows so that they can easily be examined, but also providing labels for orders, suborders, families, subfamilies (if used), and for lower categories down to species and subspecies (if used). Only specific or subspecific identification labels are attached directly to the specimen. Other labels are attached or pinned to the bottom of the container.

Killing, Preserving, and Mounting Arthropods

Order	Kill in:	Preserve in / mount:
WINGLESS ARTHROPODS All orders All stages	Hot water or alcohol*	Alcohol or prepare slide mounts
WINGED INSECTS Ephemeroptera Plecoptera	Adults—alcohol Naiads—alcohol	Alcohol
Odonata	Adults—ethyl acetate, or freezing Naiads—alcohol	Adults—pin just to the right of mid-thorax and spread wings or store in transparent envelopes. Naiads—alcohol
Blattaria, Orthoptera, Phasmida, Dermaptera	Adults—ethyl acetate or freezing Nymphs—alcohol	Adults—pin just to the right of the mid-prothorax.
Grylloblattodea, Isoptera, Zoraptera, Pscocodea, Embiidina, Thysanoptera, Strepsiptera, and all soft-bodied nymphs	Hot water or alcohol	Alcohol or prepare slide mounts
Pscocodea (ectoparasitic forms)	Hot water or alcohol	70% alcohol or clear in 10% KOH and prepare slide mounts
Hemiptera**	Adults—ethyl acetate or alcohol Nymphs—alcohol	Adults—pin through scutellum just behind mesothorax (left wings may be spread) or point. Nymphs—alcohol or slide mounts

Neuroptera, Megaloptera, Rhaphidoptera, Mecoptera	Adults—ethyl acetate, alcohol, or freezing Larvae—hot water or alcohol	Adults— pin just to the right of mid-thorax; wings may be spread. Larvae—alcohol
Lepidoptera	Adults— ethyl acetate or freezing Larvae—hot water or alcohol	Adults— pin just to the right of mid-thorax; spread left or both wings. Larvae in alcohol
Coleoptera	Adults—alcohol, ethyl acetate, or freezing Larvae—alcohol or hot water	Adults—pin though right elytron near base, or point. Larvae—alcohol
Diptera Hymenoptera	Adults— ethyl acetate or freezing Larvae—hot water or alcohol	Adults— pin just to the right of mid-thorax (wings may be spread) or point. Larvae—alcohol
Siphonaptera	Adults—alcohol Larvae—hot water or alcohol	Adults—clear in 10% KOH, prepare slide mounts Larvae—alcohol

*Alcohol should be 70 to 85%
**Scale insects and some others must be specifically prepared.

Some Special Techniques for Collecting and Preserving Insects

Thysanura and Archaeognatha are very quick-moving insects that often have a pattern of scales on the body. Catching them by conventional methods often results in rubbing off the scales and damaging the delicate legs, antennae, and cerci. Household species can be trapped in jars or bottles baited with starch or flour. Since specimens are easily rubbed or broken, handle them as little as possible. Drop into 70% alcohol and agitate as little as possible in transport.

Odonata are often difficult to collect because of the agile flight of the adults. It is often best to take up a fixed position with the net in hand and wait for the dragonflies to approach. Some specialists resort to extreme measures, such as shooting them with dust shot or with sand in a slingshot. Beating lines of individual collectors with nets, advancing along a shallow streambed, are sometimes very effective. Many smaller damselflies and dragonflies are readily collected along the edges of streams or other bodies of water. Naiads are readily collected with an aquatic dip net. Adults are best preserved dry in cellophane envelopes with the wings folded over the back, or they may be pinned through the side with the wings folded over the back and extended to the left of the pin. Fully spread adults take up a great deal of space, but they may be more aesthetically pleasing. Naiads and exuviae (shed skins) may be pinned, but are best preserved in 70% alcohol.

Ephemeroptera are usually much more easily netted than are dragonflies, but some exotic species are swift and adept fliers. Adults and subimagos can often be collected by picking them off vegetation, rocks, or bridges along suitable streams. Subimagos should be kept alive in paper bags or other containers until the provisional subimaginal skin is shed. Naiads can be collected with aquatic nets or by removing sticks, stones, or other objects from streams and watching for the creeping or scrambling naiads. Species that burrow in the bottom are often picked up with bottom-sampling devices. Adults and naiads should be preserved in 70% alcohol. If pinned adults are desired for demonstration purposes, they may be mounted on large pieces of heavy paper to protect the delicate legs, cerci, and median filament. Adults that are simply pinned become so distorted and break so easily that they are almost worthless.

Hemiptera are generally easily collected by sweeping or beating. Adults may be pinned directly if large, or mounted on points if small. Many species of leafhoppers and others, however, can only be determined to species by examination of the male genitalia. These may be prepared for study by clipping off the end of the abdomen of a pinned or pointed specimen, soaking it for a few minutes in hot 10% KOH or for several hours in cold KOH or NaOH, dehydrating in 95% and 100% alcohol, clearing in xylene (xylol), and mounting on microscope slides in Canadian balsam or synthetic medium. In some cases, the genitalia must be spread out or dissected while still in alcohol. Nymphs are generally best preserved in 70% alcohol, but larger ones may be pinned or pointed.

Scale insects (Coccoidea), jumping plant lice (Psylloidea), and aphids (Aphidoidea) may require special preparation for precise identification. Scale insects, especially the nymphs and females, often look so different from other insects that they may be easily overlooked. They are often inactive, enclosed in galls, resemble galls, or are concealed with a flattened cover (the scale), or with filaments of wax, mealy powder, waxy plates, or cottony materials, or otherwise look unlike ordinary insects. In hunting for scales, as well as aphids or psyllids, plants showing wilting or yellowing of the foliage, unusual leaf or stem growths, or uneven excrescences on the stem or bark should be carefully examined. Male scale insects are more clearly insectan, but they are rarely encountered unless reared from infestations.

In order to make precise determinations of female and nymphal scale insects, they must be prepared for study with the compound microscope. After killing, a portion of an infestation may be dried and pinned along with a portion of the leaf or stem on which it is located. Usually, most of the insects will remain attached upon drying. Such samples may also be preserved in small cellophane or wax envelopes.

Individuals from the infestation, either from fresh or dried samples, are best killed or softened in Carnoy's solution: glacial acetic acid, 1 part; absolute alcohol, 6 parts; chloroform, 3 parts; mercuric chloride to saturation. Specimens with heavy wax coverings will fix better if punctured with a needle. Fix for at least four hours before washing three times in 70% alcohol and preserving in 80% alcohol.

Temporary mounts may be prepared for quick identification or for routine examinations. Place the killed and fixed specimen on a microscope slide in a drop of acetic glycerin (glycerin 9 parts; acetic acid 1 part). Remove the scale if necessary. Boil the preparation gently after covering with a cover slip. Such temporary mounts usually allow examination of the characteristic plates, glands, spines, and lobes of the body. They may be preserved for a time by ringing the coverslip with colorless nail polish or mounting medium. Temporary mounts may also be prepared with polyvinyl alcohol mounting medium (PVA) to which a small amount of lignin pink, a chitin stain, has been added. PVA is best purchased from chemical supply dealers. Its preparation requires careful control or the results will be very unsatisfactory.

Permanent slide mounts may be prepared by placing the killed and fixed specimens in 10% KOH, dissecting out parts desired for study, and after washing in alcohol, mounting the specimen in acetic glycerin. Such slides must be carefully ringed with colorless nail polish or mounting medium and preferably should be placed in a suitable drying oven for several days to insure that the glycerin is completely water-free.

A more detailed procedure for preparing permanent mounts is as follows:

1) Kill in Carnoy's solution
Boil in 10% KOH to remove waxes
   (Add acetic acid if waxes are very thick)
Remove scale if present
Wash in distilled water twice
Dissect if necessary
Dehydrate in 95% alcohol

2) Stain in acid fushsin, magenta, or gentian violet
Destain in 95% alcohol
Dehydrate in 100% alcohol
Clear in xylene (xylol) or carbo-xylene
Mount beneath cover slip in Canadian balsam or synthetic mounting medium

2a) Dehydrate in 100% alcohol
Stain in picro-cresote (picric acid, 1 gram; beechwood creosote, 100 ml)
Destaining not necessary
Wash and clear in beechwood creosote
Mount in Canadian balsam or synthetic mounting medium

Most scales are relatively flat and can be mounted without any trouble on ordinary slides. Thick or large specimens may have to be cut around the edges and flattened. Depression slides may be useful in some cases, but visibility of characters is more important than appearance. As in all cases, only a single specimen should be mounted on each slide.

Samples of dried infestation, alcoholic specimens, and prepared mounts should be correlated with notes on the host plant, the appearance of the living colony, and any notes on observed behavior, as well as the date, locality, and collector.

Aphids and psyllids may be mounted on slides using the techniques suggested for the scale insects. It is often adequate, however, simply to kill, boil or soak in 10% KOH, wash, dehydrate, and mount in Canadian balsam or synthetic mounting medium. The chitinous structures are usually dense enough not to require special staining.

Thysanoptera and the bark lice (Pscocodea) are frequently collected in Berlese funnel collections, but can also be collected by examining flowers, bark, fungi, or debris and picking up specimens with an aspirator or forceps. A species of Thysanoptera is almost always found at the bases of the leaves of yucca plants. Alcohol specimens are usually identifiable, but slide mounts may be prepared as outlined above for the aphids and other hemiperoid insects.

The biting and sucking lice are usually found on or with their mammalian or bird hosts. Etherizing a bird or small mammal in a plastic bag often produces an abundance of lice. Injured animals or birds are particularly apt to be infested. Caged animals often develop heavy infestations. Specimens can be kept in alcohol or prepared as slide mounts by the methods outlined above for the aphids and other hemipteroids.

Blattaria and Mantodea are usually killed in ethyl acetate, alcohol, or by freezing, and are pinned with one or both pairs of wings spread in some adults. The cockroaches are usually found beneath cover or moving about at night. The mantids are often found on vegetation, where they may be difficult to detect because of their camouflaged appearance. Nymphs are best prepared in 70% alcohol.

Isoptera, Grylloblattodea, and Zoraptera are best collected into and retained in 70% alcohol. Carnoy's solution is also good for killing and fixing for detailed studies of anatomy. Dried adults usually shrivel and shrink until

they are useless, unless dehydrated in 95% and 10[...] ol, transferred to
xylene, and then dried. This technique can also be u[...] other arthropods,
such as spiders, if display or demonstration specime[...] sired.

Dermaptera are usually found under cove[...] ghts. Specimens
can be killed and preserved in 70% alcohol or h[...] nted on pins or
points. Only young nymphs are so soft that they s[...] retained in 70%
alcohol.

Orthoptera and Phasmatodea are traditiona[...] l in ethyl acetate
or by freezing and pinned. Only one pair of wing[...] ly spread, if any.
The Nearctic walking sticks are largely wingless or h[...] very short wings.
Green species will usually fade even if carefully [...] o that notes on
original colors are desirable. Soft-bodied nymph[...] be preserved in
70% alcohol.

Plecoptera and Embiidina are best killed [...] eserved in 70%
alcohol both as adults and as nymphs. Adult stone[...] ften found along
streams on stones, vegetation, or bridge abutment[...] easily be picked
up with forceps. Webspinners are sometimes foun[...] ir silken tubes in
old houses in the southern or western U.S. Some [...] so build tubes in
the ground or beneath stones.

Neuropteroid adults are usually killed in et[...] te, alcohol, or by
freezing and pinned, but they can also be killed a[...] l in 70% alcohol
satisfactorily. Larvae of these and all other holor[...] insects are best
killed in hot water or Carnoy's solution and preserv[...] o alcohol. Adult
neuropteroids are usually found while flying or at lig[...] ght.

Mecoptera are usually killed with ethyl acet[...] ol, or by freezing
and pinned. The snow scorpionflies are sometime[...] on the surface of
snow in the fall or spring of the year, but are mo[...] hern or western.
The scorpionflies are largely caught flying or some [...] ncentrated about
dead animals. *Bittacus* and its allies can often be [...] by sweeping the
shrub layer in woods.

Siphonaptera are often associated with th[...] birds or animals,
but the larvae are free-living. Fleas can often be fou[...] ogs, cats, or other
animals by combing the animal over a pan of wate[...] ol. They usually
leave the host shortly after death, so live animals ar[...] referable sources
than dead ones. Only the so-called stick-tight [...] ain permanently
attached to the host. Adults are easily preserve[...] hol and may be
prepared as microscope slide mounts by boiling or [...] in 10% KOH or
NaOH, dehydrating, immersing in xylene, and mou[...] Canadian balsam
or synthetic mounting medium. Clearing the heavy [...] from the cuticle
is more important than staining in this group. I[...] sentinel animals
such as white rats or guinea pigs can be used to coll[...] such as *Xenopsylla*
cheopis, the plague flea.

Most Coleoptera are easily collected by picking them up with the fingers or forceps and dropping them into 70% alcohol. Although usually pinned, they may be kept in 70% alcohol for up to 20 years without adverse effects. Sweeping, beating, and the use of the Berlese funnel are especially productive methods of collecting. Light traps are extremely effective in attracting beetles. Only a few fast-flying forms, such as tiger beetles, require netting.

Adult beetles killed in alcohol or ethyl acetate can be pinned or pointed directly with a minimum of drying. Care should be used in mounting so that legs, ventral structures, and other organs are not obscured in small forms. If material has been dried, it should be relaxed before attempting to pin specimens, but small specimens can be easily glued on points while quite dry. Some smaller beetles have to be dissected and mounted on slides.

Strepsipterans are usually obtained only by examining suitable hosts or by rearing adults from infected hosts. Immature stages should be killed in hot water or Carnoy's solution if at all possible and preserved in 70% alcohol. Slide preparations are often necessary for effective study of larvae.

Lepidoptera and Trichoptera adults must be netted or collected at light at night. Killing is traditionally done in an ethyl acetate jar, but adults of both butterflies and moths can be mounted from 70% alcohol. It is essential to spread one or both pairs of wings even of very small lepidopterans. Freshly killed material can be spread on a pinning board made by cutting an elongate groove in a thick piece of Styrofoam or other soft material. The wings are held down with strips of paper pinned to the board and allowed to dry. If pins are left in the wings until they dry, the holes will be quite noticeable. These usually close if the pins are removed before drying. Small moths can be spread by placing them dorsum-down on pieces of cardboard in alcohol. One or both wings can then be spread so that the color pattern and wing venation is visible. After drying, the moth can be separated from the cardboard and mounted on a point. If the left wings are spread, the specimen will be in the position of most illustrations of wing venation when examined from below. Enough of the wing venation for family determination can usually be seen without special preparation, but for detailed studies, wings mounted in balsam or other media or denuded of their scales may be needed.

Caterpillars and pupae can often be found by looking on vegetation, under bark, in leaf mold, or in soil. Immature stages are best killed with hot water or Carnoy's solution and preserved in 70% alcohol.

Hymenoptera, even very small forms, are best collected with a net and handled carefully. Large bees and wasps, which may be dangerous, can be knocked out by placing the end of the net with them in it in an ethyl acetate killing jar. They may then be transferred into the same or another jar with safety. Many parasitic forms can be reared from galls, mantid egg cases, other insects, flower heads, and many other materials. Clear plastic bags are cheap and handy containers for such rearing.

Hymenopterans are traditionally killed in ethyl acetate and mounted dry. This is especially important with very hairy bees and wasps. Immersion in alcohol causes the hairs to collapse upon the body and a very uncharacteristic appearance results. Large specimens are usually mounted directly by pinning through the thorax, and it is usually unnecessary to spread the wings except for display purposes. Small specimens can be glued to points with the left side down and the head pointing forward. This gives maximum protection for the antennae and legs. Immature stages should be killed in hot water or Carnoy's solution and preserved in 70% alcohol.

Diptera should, if possible, be killed in ethyl acetate and kept dry. Large specimens can be killed in 70% alcohol and later pinned or pointed, but small specimens usually have the wings collapse against the body and cannot be successfully mounted. Dry specimens can be mounted on points without relaxing them, providing that care is taken. Slides of male genitalia are often necessary, and can be easily prepared as described under the section regarding the Hemiptera. Whole mounts of very small flies are often useful.

Adult flies are among the most conspicuous of flying insects. They can often be netted in abundance over flowers or swarming in the air. Special searching on the bark of trees and in other places may be necessary for sand flies and some others. Adult mosquitoes are attracted to light and may be trapped, or special traps baited with dry ice (carbon dioxide) may be useful.

Immature flies are also abundant in many habitats. Larval mosquitoes occur in many types of aquatic habitats such as water-filled tree holes, water-holding plants, artificial containers such as old tires, small ponds, swamps, and even the edges of streams. Immature stages should, if possible, be killed in hot water or Carnoy's solution, although alcoholic preservation is usually adequate to identification in regions where the fauna is well known.

APPENDIX 2: IDENTIFYING THE ORDERS OF INSECTS AND NONINSECTAN HEXAPODS

A common method of determining the identification of an insect is with the use of dichotomous keys. A key is simply a tabular display of the sets of characteristics or characters defining a group. Each line, or set of lines, is called a rubric, and each pair of lines or sets of lines is called a couplet. The characters listed in one rubric compare to those listed in the companion rubric. In using a key, one is faced with a series of choices until a name is reached or one is referred to another key. Unfortunately, animals vary and absolute dichotomies are not always possible. Therefore, the user of a key must make allowances for this, and select the best fit for the specimen or specimens she or he has in hand.

To aid the user, the couplets and rubrics of the key are numbered. For example, couplet 1 of each key is composed of rubrics 1 and 1'. Each rubric is ended by a number or by the name of a taxonomic group. The end number refers the user to a couplet in the key beyond the first couplet. Read both rubrics of each couplet and decide which best fits the specimen or specimens in hand; then proceed to the couplet indicated or check the group that you have identified. In this way, by series of dichotomous choices, the user proceeds from the first couplet forward. The keys may also be used with known specimens to compile lists of characteristics of groups.

After the number of the first rubric of a couplet is a number referring to the rubric of a couplet, which referred the user to this point. This is of little value in short keys, but in longer ones, it may save valuable time by allowing quick reference to some point where an invalid decision was made. Thus, if in couplet 84 (23), the choices prove meaningless, one can quickly return to couplet 23 to reevaluate the choices there.

Some material is included in keys that is considered helpful without being diagnostic. For example, the size of insects in most families is variable, but it is often helpful to know that most members are minute insects, less than 1 mm long, or large insects, over 20 mm long, or more than 25 mm across the wings. Likewise, habitat is seldom definitive for a whole family, but the knowledge of the general habitat may limit the search. The keys are artificial in the sense that they do not follow the phylogenetic order of relationship of the families or other groups.

Beyond the Order

Knowing the order to which an insect or other arthropod belongs allows one to find a certain amount of information in the printed literature concerning entomology. For advanced work, however, the family, generic, and specific names are required. In particular, knowing the family to which an insect belongs is critical in continuing the process of identification. There are no fixed rules for attempting to identify an arthropod beyond the family. The following are only suggestions as to how one might try.

There are essentially three ways to proceed in making an identification. First, specimens may be identified by keying them out further using published books, or by comparing them with figures in a manual or textbook. Second, specimens may be carefully compared with specimens identified by some specialist in the family concerned. Finally, specimens may be sent to a specialist for identification.

In large cities, the second method of identification is often the easiest to apply. However, it teaches the student little about the insects concerned. The third method is even more restrictive. No specialist will determine miscellaneous, unmounted, unsorted lots of insects. Specimens must be correctly prepared, labeled, and sorted. The second and third methods should, therefore, follow the first and be used only after all other resources have been exhausted.

Key to the Orders of Insects and noninsectan Hexapods (Adults)

1.	Wings present..2	
1'.	Wings absent or very small (vestigial) 32	
2(1).	Forewings (wings of the mesothorax) horny, leathery, or parchment-like, at least at base; hind wings membranous (rarely lacking) often folding and concealed beneath forewings; prothorax large, not fused with mesothorax 3	
2'	One or two pairs of membranous wings (transparent or translucent) 11	
3(2).	Forewings with visible veins; hind wings folding fan-like beneath forewings or at least not folded crosswise as well as lengthwise 4	

3'.	Forewings with visible veins, uniformly horny elytra, never overlapping along midline of back	10
4(3).	Mouthparts forming a segmented beak adapted for piercing and sucking	5
4'.	Mouthparts with chewing mandibles which move laterally when in use	6
5(4).	Head usually horizontal (projecting forward) with beak projecting downward; forewings usually leathery at base and abruptly membranous toward tip (hemelytra); membranous portions of wings usually overlapping one another and lying flat over abdomen when at rest	HEMIPTERA (HETEROPTERA)
5'.	Head usually vertical with beak arising from hind part and projecting backward between legs (sometimes apparently arising from between front legs; forewings uniformly leathery (tegminous) or, if membrane is clear, with many strong longitudinal supporting veins; wings not lying flat over abdomen whether overlapping or not	HEMIPTERA (former Homoptera)
6(4').	Hind wings similar to forewings and not folded beneath them; social insects often found in colonies with wingless formsSome Tropical ISOPTERA	
6'.	Hind wings membranous, broader than forewings, folding fanlike beneath them	7
7(6').	Hind femora usually larger than fore femora, adapted for jumping, if not, forelegs forming broad, burrowing organs; forewings usually held sloping against sides of body when at rest or overlapping on back; most species capable of producing chirping or creaking sounds in male at least	ORTHOPTERA
7'.	Hind femora seldom larger than fore femora, not adapted for jumping; body form more or less flattened with wings superposed when at rest	8
8(7').	Body elongate, the head free, not concealed from above by pronotum	9

8'.   Body oval, often much flattened; head near[...] when viewed from above; legs all similar, [...] coxae large .... [...]led by pronotum [o]r rapid running, ... BLATTARIA

9(8').   Prothorax much longer than mesothorax[...] formed for seizing prey, heavily spined; [...] segments .... [le]gs nearly always [usu]ally with several ... MANTODEA

9'.   Prothorax short, usually shorter than [...] adapted for climbing or walking; .... [thor]ax; legs similar, one-segmented [PH]ASMATODEA

10(3').   Abdomen with movable, pincer-like forceps[...] slender; forewings short, leaving most of [...] wings nearly semicircular, radially folded [...] species wingless) .... [a]ntennae long and [the]n exposed; hind [ne]ar center (some DERMAPTERA

10'.   Abdomen usually covered by the conjoin[ed...] elytra are short, the abdomen is seldom (n[...] terminated by forceps; generally heavily [...] insects .... when at rest; if [A]merican species) [hard]-bodied COLEOPTERA

11(2').   Two pairs of wings present (four-winged) 15

11'.   Only one pair of membranous wings presen[t] 12

12(11').   Small, active, very rare insects with forewing[s...] the hind wings membranous and without c[...] or 5-segmented, with or without claws; an[...] lateral process, no cerci................................. [reduc]ed to little straps, [...]; tarsi 2-, 3-, 4-, [sh]ort, with a long STREPSIPTERA

12'.   Forewings developed, the hind wings reduce[d...] [...]ing 13

13(12').   Forewings milky or infuscated with numero[us...] few cross-veins; mouthparts reduced (ve[stigial in] adult; antennae short, filiform; abdomen [with] segments and terminated by two elongate[...] filaments and often a long median filam[ent...] segmented; wing spread about 10 mm or les[s]............................ EPHEMEROPTE[RA] [longit]udinal veins, but [n]on-functional in [t]apering, with 10 [s]egmented caudal of hind legs 4- [...] [famil]y Caenidae only)

13'   Wings variable, but hind wings usually detectable as small hook-like slips or distinct knob-like halteres; abdomen with fewer than 10 segments, never with elongate caudal filaments..................14

14(13'). Small insects with long antennae (10-25 segments), forewings with greatly reduced venation; hind wings not forming knob-like halteres; mouthparts lacking; abdomen terminated by a single, elongate style-like process; tarsi 1-segmented..
....Males of scale insects, HEMIPTERA

14'.   Minute to moderately large insects with hind wings forming distinct knob-like halteres; mouthparts modified for sucking, piercing, or lapping, rarely reduced or absent; antennae variable either elongate with 6 or more segments or short with only 3 segments, the last of which may be secondarily annulated; tarsi 5-segmented..... DIPTERA

15(11'). Wings long, narrow, almost without veins, fringed by relatively long hairs; tarsi 1- or 2-segmented, swollen at tip without claws; mouthparts asymmetrical, without mandibles, adapted for lacerating and sucking; most species minute........................THYSANOPTERA

15'.   Wings broader and usually with some veins; tarsi with more than two segments, the last not swollen, and usually with claws.......................16

16(15'). Hind wings large, larger than forewings, and with an anal area which folds fan-like in repose; wing veins usually numerous; antennae usually long and many-segmented..17

16'.   Hind wings about same size as forewings or smaller, without an anal area which folds fan-like..19

17(16'). Tarsi 5-segmented; cerci not prominent..18

17'.   Tarsi 3-segmented; body usually somewhat flattened, with prominent segmented cerci at tip; wings at rest overlapping on back; prothorax large, free; species usually of moderate size.................PLECOPTERA

18(17'). Costal area with few cross-veins (area between first and second veins at anterior edge of wing); wing membranes hairy with many tactile setae; prothorax small; small to moderately large insects......................
.... TRICHOPTERA

18'. Costal area with many cross-veins; prothorax large; moderate to very large insects .... MEGALOPTERA

19(16'). Antennae short, inconspicuous; wings net-veined with many cross-veins 20

19'. Antennae short, more conspicuous, wings with few cross-veins 21

20(19'). Hind wings much smaller than forewings (sometimes lacking, see couplet 13); abdomen terminated by long thread-like caudal processes; tarsi usually 4- or 5-segmented .... EPHEMEROPTERA

20'. Hind wings much like forewings, sometimes broader, never very small or lacking; no elongate filaments at tip of abdomen; tarsi 3-segmented .... ODONATA

21(19'). Head produced into an elongate beak with chewing mouthparts at tip; wings often with a colored pattern; male abdomen usually terminated by swollen genitalia sometimes resembling a scorpion's sting .... MECOPTERA

21'. Head not produced into an elongate beak; male genitalia not usually conspicuous 22

22(21'). Mouthparts with chewing mandibles 23

22'. Mouthparts adapted for sucking, no chewing mandibles 30

23(22). Tarsi with 5 segments (if rarely only 3- or 4-segmented, the hind wings are smaller than forewings and wings lie flat over back); no cerci 24

23'. Tarsi 2-, 3-, or 4-segmented; wings with few veins 27

24(23). Prothorax small or moderately long (if very long, the forelegs are adapted for grasping prey) 25

24'. Prothorax very long, cylindrical, much longer than head; forelegs adapted for walking; antennae with more than 11 segments; wings with numerous veins and cross-veins .... RHAPHIDIOPTERA

25(24). Wings similar with many veins and cross-veins (if rarely the venation is reduced, wings are covered with whitish powder) 26

25'.	Wings with relatively few veins and cross-veins, sometimes almost veinless; costal cell of forewing without cross-veins; hind wings smaller than forewings; prothorax with mesothorax; abdomen often constricted at base and terminating in a sting or specialized ovipositor .... HYMENOPTERA
26(25).	Costal cell in forewing almost always with many cross-veins .... NEUROPTERA
26'.	Costal cell without cross-veins; mouthparts on a prolonged beak .... MECOPTERA
27(23').	Fore and hind wings nearly equal in size and with similar venation; wings held more or less flat, superposed on back when at rest; tarsi 3-, 4-, or 5-segmented 28
27'.	Hind wings smaller than forewings; wings folded back against abdomen when at rest; tarsi 2- or 3-segmented 29
28(27).	Tarsi apparently 4-segmented; wings often elaborately net-veined; wings shed after mating flight; cerci usually minute; social insects often found associated in colonies with wingless individuals .... ISOPTERA
28'.	Tarsi 3-segmented; basal segment of forelegs enlarged, bulb-like; cerci conspicuous; only male is winged; wings with few veins .... EMBIIDINA
29(27').	Cerci lacking; tarsi 2- or 3-segmented; wings permanently attached, wings often with a characteristic zig-zag venation with few cross-veins, forewings sometimes thickened.. .... PSOCODEA (former Psocoptera)
29'.	Cerci present, conspicuous; tarsi 2-segmented; wings with greatly reduced venation, shed after mating flight .... ZORAPTERA
30(22').	Wings not covered with flattened scales, antennae with five or fewer segments, mouthparts present and thick in appearance 31
30'.	Wings and body covered with flattened, colored scales (modified setae) which usually form definite patterns on the wings; antennae usually long, many-segmented; mouthparts (sometimes absent)

usually in the form of an elongate sucking tube, spirally coiled beneath head when at rest..LEPIDOPTERA

31(30). Sucking beak apparently arising from front of head.......HEMIPTERA (HETEROPTERA)

31'. Sucking beak apparently arising from back of head................................
....HEMIPTERA (former Homoptera)

WINGLESS OR WITH GREATLY REDUCED WINGS

32(1'). Body insect-like (with distinct head, thorax, and abdomen) and usually six segmented legs on thorax..33

32'. Body more or less flattened, legless, with an elongate sucking beak; usually covered with a waxy scale, or powdery or with cottony tufts; found feeding on plants often in groups ....
....HEMIPTERA (former Homoptera)

33(32). Body strongly laterally compressed (flattened); mouthparts forming a sharp, inflexed beak; hind femora enlarged, adapted for jumping; adults parasites on birds and mammals................... SIPHONAPTERA

33'. Body not strongly laterally compressed, of various forms..................34

34(33'). Body dorsoventrally compressed..35

34'. Body not strongly dorsoventrally compressed......................................41

35(34). Mouthparts adapted for chewing..36

35'. Mouthparts adapted for piercing and sucking (haustellate)................39

36(35). Body flattened and oval; vestigial wings sometimes present; legs adapted for running; pronotum nearly conceals head; antennae usually long and slender..Some BLATTARIA

36'. Pronotum not concealing head, antennae usually short; legs not well adapted for running...37

37(36'). Cerci on abdomen long 38

37'. No cerci; generally elongate oval insects with triangular heads; parasites of birds and mammals..
.... PSOCODEA (former Phthiraptera)

38(37). Cerci simple, straight; eyes absent; antennae short; external parasites of African rodents..
.... DERMAPTERA (suborder Diploglossata or family Hemimeridae)

38'. Cerci not segmented, bent or angulate at middle, resembling forceps, eyes present; antennae nearly as long as body; associates of bats in bat caves in southern Asia..
.... Family Arixendae of DERMAPTERA

39(35'). Antennae visible, though rather short..40

39'. Antennae inserted in pits, not visible from above; legs usually rather long; hairy, wingless flies often parasitic on or associated with birds, mammals, or in bee hives................................... Pupiparous DIPTERA

40(39). Beak unsegmented; eyes reduced or absent; tarsi forming hooks for grasping hairs of host; parasites on mammals only................................
.... PSOCODEA (former Phthiraptera)

40'. Beak segmented; eyes present or absent; tarsi not forming hooks; vestigial wings often detectable on thorax; flattened hairy bugs often parasitizing or associated with birds or mammals, or in termite colonies. ....Some HEMIPTERA (HETEROPTERA)

41(34'). Mouthparts small, surrounded by outgrowths of the sides of head, not usually visible from without; underside of abdomen with styles, small leg-like appendages, or other appendages 42

41'. Mountparts usually fully visible, adapted for chewing or sucking.....44

42(41). Minute, cylindrical hexapods often with the head pear-shaped, pointed anteriorly; forelegs used as substitutes for antennae, eyes, and cerci, lacking in life; basal three abdominal segments with small leg-like appendages or styles, abdomen in adult 12-segmented, terminated by a small anal tube..PROTURA

42'. Antennae present although eyes and cerci may be lacking; head not pear-shaped or pointed..43

43(42'). Abdomen with only six segments or less, … segment below and often with a spring (…); cerci absent; eyes simple, similar to eyes of … present in small groups at sides of head ……… … sucker on basal … near tip beneath; … larvae, sometimes … COLLEMBOLA

43'. Abdomen with 9 to 11 segments; antennae … bead-like segments (moniliform); eyes … elongate, segmented feelers or modified … palpi of mouthparts exposed; usually co… found in soil or under rocks and debris ……… … gate, often with … cerci present as … cer-like forceps; white hexapods ……… DIPLURA

44(41'). Body covered with minute, flattened scales … at least tips of mandibles and other … exposed; antennae nearly always long and … a median abdominal filament also usually … ……………………………………………………………… … pted for running; … ts visible; palpi … nented; cerci and many-segmented …………………45

44'. Body not covered with scales or, if so, … beneath the head……………………………………… … ubular proboscis …………………46

45(44). Eyes large, composed of many minute facets … over the front and usually touching each o… … very long, 7-segmented; four posterior … hooked processes which are also present … segments (vestiges of legs); body usually str… ……………………………………………………………… … atidia), extending … e; maxillary palpi … basal styles or … of the abdominal … vex above ……… MAEOGNATHA

45'. Eyes small or lacking, when present comp… set at sides of head and not extending ove… or 6-segmented; 11th abdominal tergite par… and abdomen without styles; body often fla… … only a few facets … maxillary palpi 5-… ed by 10th; coxae … …THYSANURA

46(44'). Mouthparts distinctly formed for biting or … ………………………47

46'. Mouthparts adapted for sucking, forming a … roboscis ………59

47(46'). Abdomen terminated by strong, move… pronotum free……………………………………… cer-like forceps; DERMAPTERA

47'. Abdomen not ending in forceps……………… ………………48

48(47'). Abdomen not strongly constricted at base ……………………………………………………………… joined to thorax …………………49

48'.	Abdomen strongly constricted at base (wasp-waisted); prothorax fused with mesothorax ....Some HYMENOPTERA (ants, velvet ants, etc.)
49(48).	Head not prolonged into a beak 50
49'.	Head prolonged into a beak with chewing mouthparts at tip .... Some MECOPTERA
50(49).	Very small (less than 3 mm long) jumping insects; prothorax very small, inconspicuous ....Some PSOCOPTERA
50'.	Prothorax usually large; size usually larger 51
51(50').	Hind legs adapted for jumping, the femora large................................... .... Some ORTHOPTERA
51'.	Hind legs not adapted for jumping...52
52(51').	Prothorax much longer than mesothorax; front legs nearly always adapted for grasping prey with many spines .... Some MANTODEA
52'.	Prothorax not greatly lengthened..53
53(52').	Cerci present; antennae usually present with more than 15 segments, often many-segmented 54
53'.	Cerci absent; body often very hard-shelled; antennae usually 11-segmented..Some COLEOPTERA
54(53).	Cerci with more than 3 segments 55
54'.	Cerci short, with 1 to 3 segments 57
55(54)	Body more or less flattened, oval; head almost or quite concealed by pronotum; legs adapted for running; antennae usually long, many-segmented..Some BLATTARIA
55'.	Body elongate; head nearly horizontal, exposed....................................56

56(55'). Cerci long; abdomen terminated in female by an exerted, sclerotized ovipositor; tarsi 5-segmented; rare insects found in mountains of Western N.A. and Japan..................................GRYLLOBLATTODEA

56'. Cerci short; no exposed ovipositor; tarsi 4-segmented.................Some ISOPTERA

57(54'). Tarsi 2- to 4-segmented; body not greatly elongate..............................58

57'. Tarsi 5-segmented; body long and slender...64

58(57'). Front tarsi not enlarged...59

58'. Basal segment of fore tarsus enlarged, bulb-like in both nymphs and adults; females always wingless, living in small colonies with nymphs and dealate males in silken tunnels spin by silk glands in fore tarsi ....EMBIIDINA

59(58). Tarsi apparently 4-segmented; cerci with several segments; antennae with 9 to 30 segments...Some ISOPTERA

59'. Tarsi 2-segmented; cerci 1-segmented; antennae 9-segmented; usually small insects living in small colonies of nymphs, wingless adults, and dealate adults...ZORAPTERA

60(46'). Body bare or with scattered hairs or a waxy coating.........................61

60'. Body densely clothed with flattened scales and hairs; proboscis usually present, appearing as a coiled tube beneath the head ....LEPIDOPTERA

61(60). Tarsi usually simple, with distinct claws..62

61'. Tarsi with last joint swollen, bladder-like; proboscis a triangular unsegmented beak; small insects often in leaf mold or on vegetation ....Some THYSANOPTERA

62(60). Prothorax distinct..63

62'. Prothorax small, hidden when viewed from above..Some DIPTERA

63(62). Beak arising from front of head... ....HEMIPTERA (HETEROPTERA)

63'.   Beak arising from back part of head ....
         ....HEMIPTERA(former Homoptera)

64(57)   Mouthparts hypognathous, pretarsi raised while walking; known only from southwestern Africa .... MANTOPHASMATODEA

64'   Mouthparts prognathous, pretarsi not raised while walking...................
         ....many PHASMATODEA

ADDITIONAL READINGS

Arnett, R.H. Jr. 2000. American insects: a handbook of the insects of America north of Mexico. Boca Raton, FL, CRC Press.

Berenbaum, M.R. 1995. Bugs in the system: insects and their impact on human affairs. Reading, MA; Addison Wesley.

Borror, D.J. and R.E. White. 1998. A field guide to insects. Boston, Houghton Mifflin Harcourt.

Chapman, R.F. 1998. The Insects: structure and function. Fourth Edition. Cambridge, Cambridge University Press.

Eaton, E.R. and K. Kaufman. 2007. Kaufman field guide to insects of North America. New York, Houghton Mifflin Co.

Grimaldi, D. and M.S. Engel. 2005. Evolution of the Insects. Cambridge, Cambridge University Press.

Gullan, P.J. and P.S. Cranston. 2000. The insects: an outline of entomology. Second Edition. Oxford, Blackwell Science.

Kritsky, G. 2010. The Quest for the perfect hive. Oxford, Oxford University Press.

Kritsky, G. and R. Cherry. 2000. Insect mythology. Lincoln, NE, Writer's Club Press.

McGavin, G. C. 2001. Essential entomology. Oxford, Oxford University Press.

Opler, P.A. 1998. Eastern butterflies. Boston, Houghton Mifflin Harcourt.

Opler, P.A. 1999. Western butterflies. Boston, Houghton Mifflin Harcourt.

Triplehorn, C.A. and N.F. Johnson. 2005. Borror and Delong's introduction to the study of insects. Seventh Edition. Belmont, CA, Thomsom Brooks/Cole.

White, R.E. 1998. Beetles: a field guide to beetles of North America. Boston, Houghton Mifflin Harcourt.

ABOUT THE AUTHORS

Gene Kritsky, Ph.D. (University of Illinois), is Professor of Biology and Chair of the Department of Biology at the College of Mount St. Joseph in Cincinnati, Ohio. He is also Adjunct Curator of Entomology at the Cincinnati Museum Center, Adjunct Assistant Professor of Horticulture at the University of Cincinnati, and Editor-in-Chief of *American Entomologist*. He is the author six books and over 150 scientific papers on entomology, the history of biology, and Egyptology. Dr. Kritsky has served as President of the Indiana Academy of Science and of the National Association of Academies of Science, and he was elected a Fellow of the American Association for the Advancement of Science in 1996. He was a Fulbright Scholar to Egypt in 1981-82, and a member of a scientific exchange to the Peoples' Republic of China in 1987. During 2001-2002, he transcribed Darwin's research notes for *The Descent of Man* while on sabbatical with the Darwin Correspondence Project at Cambridge University. Dr. Kritsky's research has attracted national attention with appearances on the *ABC Evening News*, *CBS Evening News*, the *Today Show*, *CBS Sunday Morning*, and *CNN*. His research has also been featured in the *U.S. News and World Report*, *USA Today*, *Parade*, *People*, *Discover*, *Scientific American*, *The New York Times*, *The Washington Post*, *Science News*, *The Scientist*, and many international publications.

Frank N. Young, Jr., Ph.D. (University of Florida), became Professor Emeritus at Indiana University in 1986, after a distinguished career as a Professor in the Department of Biological Sciences and the United States Army Medical Services Corps. He had an extensive publication record, including *The Water Beetles of Florida* and papers on herpetology, malacology, ecology, general natural history, and medical entomology, as well as general entomology. His research focused on the taxonomy and ecology of water beetles and the distributional history of periodical cicadas. He received numerous grants and awards during his career, including a Guggenheim Fellowship to the British Museum (Natural History) in London, and he was elected an E.S. George Fellow at the University of Michigan, a Fellow of the Indiana Academy of Science, and a Louisiana State University Fellow in Tropical Medicine. Dr. Young was known as a popular teacher with a keen sense of humor. He passed away in 1998.

GLOSSARY

Abdomen The hindmost of the three main body d... f an insect.

Acetylcholine A substance present in many parts c... ly of animals and important to the function of nerves.

Aedeagus The part of the male genitalia which is i... to the female during copulation and which carries the sperm into ... e. Its shape is often important in separating closely related species.

Alate Winged; having wings.

Alternating generations When two generations ar... ed within a life cycle, each producing individuals of only one sex, ei... first and then female, or vice versa.

Ametabolous Development without any noticeabl... orphosis, as found in the Protura, Diplura, Collembola, Archaeo... nd Thysanura.

Anal Pertaining to last abdominal segment which b... nus.

Anal fold A fold in the inner margin of the hindwi...

Anal veins The hindmost or most posterior longitu... ng veins.

Antenna (pl., antennae) Pair of segmented appen... ated on the head and usually sensory in function; the "feelers."

Anterior Concerning or facing the front, towards th...

Anus The posterior opening of the digestive tract.

Aorta The anterior, non-chambered, narrow part o... ct heart which opens into the head.

Appendage Any limb or other organ, such as an a... hich is attached to the body by a joint.

Apterous Without wings.

Apterygote Any member of the primitively wingle...

Arachnida A class of arthropods which include the scorpions, spiders, mites, and ticks, among others.

Archedictyon The dense arrangement of wing veins in mayflies and dragonflies.

Arista A bristle-like outgrowth from the antenna in various flies.

Aristate Bearing an arista or bristle.

Arolium A small pad between the claws on an insect's foot. Usually very small, but well developed in grasshoppers and some other insects.

Arthropoda A phylum of animals with a segmented body, exoskeleton, and jointed legs.

Axon The process of a nerve cell that conducts impulses away from the cell body.

Brachypterous With short wings that do not cover the abdomen, used of individuals of a species which otherwise has longer wings.

Caecum (pl., caeca) A sac or tubelike structure open at only one end.

Capitate With an apical knob-like enlargement.

Cardo The basal segment of the maxilla.

Carnivorous Preying or feeding on animals.

Castes Groups of individuals that become irreversibly behaviorally distinct at some point prior to reproductive maturity. One of the distinct forms that make up the population among social insects. The usual three castes are queen, drone (male), and worker. The termites and some of the ants have one or more soldier castes as well.

Caterpillar The larva of a moth, butterfly, or sawfly.

Cell An area of the wing bounded by a number of veins.

Cement layer A thin layer on the surface of insect cuticles formed by the hardened secretion of the dermal glands.

Cephalothorax A body region consisting of head and thoracic segments, as in spiders.

Cerci (singular: cercus) The paired appendages, often very long, which spring from the tip of the abdomen in many insects.

Cervical Concerning the neck region, just behind the head.

Chelicera (pl., chelicerae) The anterior pair of appendages in arachnids; the fangs.

Chitin The tough horny material, chemically known as a nitrogenous polysaccharide, which makes up the bulk of the insect cuticle; also occurs in other arthropods.

Chorion The inner shell or covering of the insect egg.

Chrysalis The pupa of a butterfly.

Cibarium The region between the hypopharynx and the clypeus. The clypeus is often enlarged with muscles that when contracted create negative pressure, drawing fluid into the mouth.

Class A division of the animal kingdom lower than a phylum and higher than an order; for example, the class Insecta.

Clavate Club-shaped, with the distal end swollen: most often applied to antennae.

Cline A progressive, usually continuous change in one or more characters of a species over a geographic or altitudinal range.

Club The thickened terminal (farthest from the head) end of the antenna.

Clypeus Lowest part of the insect face, just above the labrum.

Cocoon A case, made partly or completely of silk, which protects the pupa in many insects, especially the moths. The cocoon is made by the larva before it pupates.

Colony Of social insects, a group which co-operates in the construction of a nest and in the rearing of the young.

Complete metamorphosis Metamorphosis in which the insect develops through four distinct stages, e.g., ova or egg, larva, pupa, and adult; the wings (when present) develop internally during the larval stage. Also called holometabolous development.

Compound eye An eye consisting of many individual elements or ommatidia, each of which is represented externally by a facet.

Corpora allata A pair of small endocrine glands located just behind the brain.

Cosmopolitan Occurring throughout most of the world.

Costa One of the major longitudinal veins, usually forming the front margin of the wing and usually abbreviated to C.

Coxa The basal segment of the insect leg, often immovably attached to the body.

Crochets Hooked spines at tip of the prolegs of lepidopterous larvae.

Crop The dilated section of the foregut just behind the esophagus.

Cross-vein A short vein joining any two neighboring longitudinal veins.

Cryptic Coloring and patternation adapted for the purpose of protection from predators or prey by concealment.

Cubitus One of the major longitudinal veins, situated in the rear half of the wing and usually with 2 or 3 branches: abbreviated to Cu.

Cursorial Adapted for running.

Cuticle The outer noncellular layers of the insect integument, secreted by the epidermis.

Diaphragm A horizontal membranous partition of the body cavity.

Dicondylic Having two articulations. Most insects have dicondylic mandibles.

Dimorphism A difference in size, form, or color between individuals of the same species or the sexes within a species, characterizing two distinct types.

Dorsal On or concerning the back or top of an animal.

Dorsal ocellus The simple eye in adult insects and in nymphs and naiads.

Ecdysis The molting process, by which a young insect sheds its outer skin or pupal case.

Eclosion Emergence of the larva from the egg or of the adult or imago from the pupa.

Ectoderm The outer embryological layer which gives rise to the nervous system, integument, and several other parts of an insect.

Ectoparasite A parasite that lives on the outside of its host.

Elytron (plural elytra) The tough, horny forewing of a beetle or an earwig.

Empodium An outgrowth between the claws of a fly's foot: it may be bristle-like.

Endocuticle The innermost layer of the cuticle.

Endopterygote Any insect in which the wings develop inside the body of the early stages and in which there is a complete metamorphosis and pupal stage.

Epicuticle The thin, non-chitinous surface layers of the cuticle.

Epidermis The cellular layer of the integument that secretes or deposits a comparatively thick cuticle on its outer surface.

Epimeron The posterior part of the sidewall of any of the three thoracic segments.

Episternum The anterior part of the sidewall of any of the three thoracic segments.

Epithelium The layer of cells that covers a surface or lines a cavity.

Esophagus The narrow part of the alimentary canal immediately posterior to the pharynx and mouth.

Excretion The elimination of the waste products of metabolism.

Exocuticle The hard and usually darkened layer of the cuticle lying between the endocuticle and epicuticle.

Exoskeleton Collectively, the external plates of the body wall.

Exopterygote Any insect in which the wings develop gradually on the outside of the body, in which there is only a partial metamorphosis and no pupal stage.

Exuvia The cast-off outer skin of an insect or other arthropod.

Family A taxonomic subdivision of an order, suborder, or superfamily that contains a group of related subfamilies, tribes and genera. Family names always end in -idae.

Femur The 3rd (counting out from the body) and often the largest segment of the insect leg.

Filament A thread-like structure, especially one at the end of an antenna.

Filiform Thread-like or hair-like, applied especially to antennae.

Flabellate With projecting flaps on one side, applied especially to antennae.

Flagellum The distal (furthest away from the body) part of the antenna, beyond the 2nd segment.

Foregut The anterior part of the alimentary canal from the mouth to the midgut.

Fossorial Adapted for digging.

Frenulum The wing-coupling mechanism found in many moths.

Frons Upper part of the insect face, between and below the antennae and usually carrying the median ocellus or simple eye.

Furcula The forked spring of a springtail.

Galea The outer branch of the maxilla, the inner one the lacinia.

Gall An abnormal growth of a plant caused by the presence in its tissues of a young insect or some other organism. Aphids, gall wasps, and gall midges are among the major gall-causing insects.

Ganglion A nerve mass that serves as a center of nervous influence.

Gastric caeca The sac-like diverticula at the anterior end of the midgut.

Gena The cheek; that part of the head below and behind the eye.

Geniculate Abruptly bent or elbowed, applied especially to antennae.

Genitalia The copulatory organs of insects and other animals. The shape and arrangement of the genitalia are often used to distinguish closely related and otherwise very similar species.

Genus (pl., genera) A group of closely related species.

Gill Breathing organ possessed by many aquatic creatures, including numerous young insects. Insect gills are usually very thin growths from the body and they contain numerous air tubes, or tracheae. Oxygen passes into the tubes from the water by diffusion.

Glossa (pl., glossae) One of a pair of lobes at the tip of the labium or lower lip: usually very small, but long in honey bees and bumble bees, in which the two glossae are used to suck up nectar.

Hemolymph The blood plasma or liquid part of the blood, though generally synonymous for blood of insects.

Haltere One of the club-shaped balancing organs found on each side of the metathorax among the true flies (Diptera)

Heart The chambered, pulsatile portion of the dorsal vessel.

Head The anterior body region of insects which bears the mouthparts, eyes, and antennae.

Hemelytron (pl., hemelytra) The forewing of a hemipteran bug, differing from the beetle elytron in having the distal portion membranous.

Hemimetabolous An aquatic insect having an incomplete metamorphosis, with no pupal stage in the life history.

Hindgut The posterior part of the alimentary canal between the midgut and anus.

Homologous Organs or parts which evolved from a common ancestral organ or part, but not necessarily similarity of function.

Holometabolous Having a complete metamorphosis, with larval and pupal stages in the life history.

Hormone A chemical substance formed in some organ of the body, secreted directly into the blood, and carried to another organ or tissue where it produces a specific response.

Host The organism in or on which a parasite lives; the plant on which an insect or other arthropod feeds.

Hypopharynx A component of the insect mouthparts arising behind the mouth and just in front of the labium or lower lip.

Insecta A class of the phylum Arthropoda, distinguished by adults having three body regions: head, thorax, and abdomen; the head with a complete tentorium, and by having the thorax three-segmented with each segment bearing a pair of legs.

Instar The stage in an insect's life history between any two molts.

Integument The insect's outer coat; synonymous with the exoskeleton.

Jugum A narrow lobe projecting from the base of the forewing in certain moths and overlapping the hind wing, thereby coupling the two wings together.

Labellum The expanded tip of the labium, used by many flies to mop up surface fluids.

Labial Concerning the labium.

Labial palpus (pl., labial palpi) One of the pair of sensory appendages (feeler-like and 2 to 5 segments long) of the insect labium.

Labium The "lower lip" of the insect mouthparts, formed by the fusion of two maxilla-like appendages.

Labrum The "upper lip" of the insect mouthparts: not a true appendage, but a movable sclerite on the front of the head.

Lacinia The inner branch of the maxilla, the outer one being the galea.

Lamella (pl., lamellae) Leaflike plate.

Lamellate Possessing lamellae: an antenna with leaflike plates or extensions of antennal segments.

Larva (pl., larvae) A young insect which is markedly different from the adult: caterpillars and fly maggots are good examples.

Lateral Concerning the sides.

Lateral ocellus The simple eye in holometabolous larvae. Also called stemma (pl., stemmata).

Ligulae Name given to the lobes at the tip of the labium: usually divided into glossae and paraglossae.

Maggot A vermiform larva; a larva without legs and without well-developed head capsule.

Malpighian tubules Excretory tubes arising from the anterior end of the hindgut and extending into the body cavity.

Mandible The jaw of an insect. It may be sharply toothed and used for biting, as in grasshoppers and wasps, or it may be drawn out to form a slender needle, as in mosquitoes.

Mandibulate Having mandibles suited for biting and chewing.

Maxilla (pl., maxillae) One of the two components of the insect's mouthparts lying just behind the jaws. They assist with the detection and manipulation of food and are often drawn out into tubular structures for sucking up liquids.

Maxillary Concerning or to do with the maxillae.

Membranous Thin and transparent (in reference to wings); thin and pliable (in reference to integument).

Mesonotum The dorsal surface of the 2nd thoracic segment, the mesothorax: usually the largest thoracic sclerite.

Mesopleuron The sclerite or sclerites making up the side wall of the mesothorax.

Mesoscutum The middle and usually the largest division of the mesonotum.

Mesosternum The ventral surface or sclerite of the mesothorax.

Mesothorax The 2nd segment of the thorax.

Metamorphosis The process through which a young insect changes into its adult form.

Metanotum The dorsal surface of the metathorax. It is often very small and its subdivisions are usually obscured.

Metapleuron The sclerite or sclerites making up the side wall of the metathorax.

Metasternum The ventral surface or sclerite of the metathorax.

Metathorax The 3rd and last segment of the thorax.

Micropyle A minute opening or group of openings into the insect egg through which the spermatozoa enter in fertilization.

Midgut The middle part of the alimentary canal and the main site of digestion and absorption.

Moniliform (of antennae) Composed of bead-like segments, each well separated from the next.

Monocondylic Having only one articulation. Archaeognatha have monocondylic mandibles.

Morphology The physical structure of an organism or any of its parts.

Naiad An aquatic, gill-bearing nymph.

Nasute A type of soldier caste in certain termites. This form bears a median frontal rostrum through which it ejects a defensive fluid; the jaws are small or vestigial.

Nocturnal Active at night.

Nodus The kink or notch on the costal margin of the dragonfly wing.

Notum The dorsal or upper surface of any thoracic segment: usually prefixed by pro-, meso-, or meta- to indicate the relevant segment.

Nymph Name given to the young stages of those insects which undergo a partial metamorphosis. The nymph is usually quite similar to the adult except that its wings are not fully developed.

Occipital suture A groove running around the posterior region of the head of some insects and separating the vertex from the occiput.

Ocellus (pl., ocelli) One of the simple eyes of insects, usually occurring in a group of three on the top of the head, although one or more may be absent from many insects.

Ommatidium (pl., ommatidia) One of the units which make up the compound eyes of arthropods.

Ootheca (pl., oothecae) An egg case formed by the secretions of accessory genital glands or oviducts, such as the purse-like structure carried around by cockroaches or the spongy mass in which mantids lay their eggs.

Order A subdivision of a class or subclass containing a group of related families.

Ovipositor The tubular or valved egg-laying apparatus of a female insect; concealed in many insects, but extremely large among the bush-crickets and some parasitic hymenopterans.

Palp A segmented leg-like structure arising on the maxilla or labium. Palps have a sensory function and play a major role in tasting food.

Paraglossa One of a pair of lobes at the outer edges of the tip of the labium; with the central glossae, the paraglossae make up the ligula.

Parthenogenesis A form of reproduction in which eggs develop normally without being fertilized.

Paurometabolous Development of terrestrial insects that do not have the pupal stage.

Pectinate Having branches which arise from the main axis like the teeth of a comb: usually applied to antennae.

Pedicel The 2nd antennal segment: the name is also given to the narrow waist of an ant.

Pharynx The anterior part of the foregut between the mouth and the esophagus.

Pheromone A substance secreted by an animal which when released externally in small amounts causes a specific reaction, such as stimulation to mate with or supply food to a receiving individual of the same species.

Phylum (pl., phyla) A major division of the animal kingdom which contains various classes, orders, etc.

Pleural suture A vertical or diagonal groove on each of the thoracic pleura which separates the episternum at the front from the epimeron at the back.

Pleuron The side wall of a thoracic segment.

Plumose With numerous feathery branches; applied especially to antennae.

Pollen The mass of microspores or male fertilizing elements of flowering plants.

Pollen basket The pollen-carrying region on the hind leg of a bee: also known as the corbicula.

Posterior Concerning or facing the rear.

Postmentum The basal region of the labium.

Predaceous Preying on other animals.

Predator An animal that attacks and feeds on other animals usually smaller and weaker than itself.

Prementum The distal region of the labium, from which arise the labial palps and the ligula.

Pretarsus In insects the terminal segment of the leg bearing the pretarsal claws.

Proboscis Collective term for insect mouthparts that are drawn out to form a sucking tube.

Prognathous Having a more or less horizontal head with the mouthparts at the front.

Proleg One of the fleshy, stumpy legs on the hind end of a caterpillar.

Pronotum The dorsal surface or sclerite of the first thoracic segment.

Prothoracic gland One of a pair of endocrine glands situated in the prothorax near the prothoracic spiracles.

Prothorax The 1st or anterior thoracic segment.

Proventriculus The posterior section of the foregut.

Pterostigma A small colored area near the wingtip of dragonflies, bees, and various other clear-winged insects: also called the stigma.

Pterygote Any member of the sub-class Pterygota, which includes all winged and some secondarily wingless insects.

Pupa (pl., pupae) The third stage in the life history of butterflies and other insects undergoing a complete metamorphosis during which the larval body is rebuilt into that of the adult insect: a non-feeding and inactive stage.

Puparium (pl., puparia) The barrel-shaped case which conceals the pupa of the house fly and many other true flies. It is formed from the skin of the last larval instar.

Pupate To turn into and exist as a pupa.

Radial Sector The posterior of the two main branches of the radius, usually abbreviated to Rs.

Radius One of the main longitudinal veins, running near the front of the wing and usually the 3rd and abbreviated to R. It gives off a posterior branch (the radial sector) and the smaller branches of these veins are numbered R1, R2, etc.

Raptorial Adapted for seizing and grasping prey, like the front legs of a mantis.

Rectum In insects, the posterior expanded part of the hindgut, typically pear-shaped.

Reproductives In termites, the caste of queens and kings, in other social insects, only the queens.

Rostrum A beak or snout, applied especially to the piercing mouthparts of bugs and the elongated snouts of weevils.

Salivary glands Glands that open into the mouth and secrete a fluid with digestive, irritant, or anticoagulatory properties.

Scale A scale insect; a member of the order Hemiptera.

Scape The first antennal segment.

Sclerite Any of the individual hardened plates which make up the exoskeleton.

Sclerotization The hardening and darkening processes of the cuticle (involves the epicuticle and exocuticle with a substance called sclerotin).

Scutellum The 3rd of the major divisions of the dorsal surface of a thoracic segment: usually obvious only in the mesothorax and very large in some bugs.

Segment One of the rings or divisions of the body, or one of the sections of a jointed appendage.

Segmentation The embryological process by which the insect body becomes divided into a series of parts or segments.

Serrate Toothed like a saw.

Seta (pl., setae) A bristle.

Setaceous Bristle-like, applied especially to antennae.

Simple eye An ocellus.

Simple metamorphosis Metamorphosis in which the wings (when present) develop externally during the immature stage and there is no prolonged resting stage (i.e pupa) preceding the last molt; stages included are the egg, nymph, and adult.

Soldier In termites, sterile males or females with large heads and mandibles; they function to protect the colony.

Species The basic unit of living things, consisting of a group of individuals which can potentially interbreed with each other to produce another generation.

Spermatheca A small sac-like branch of the female reproductive tract of arthropods in which sperm may be stored.

Spermatophore A packet of sperm.

Spine A multicellular, thorn-like process or outgrowth of the integument not separated from it by a joint.

Spiracle One of the openings of the tracheal system through which diffusion of gases takes place.

Spur A large and usually movable spine, normally found on the legs.

Stage A distinct, sharply differentiated period in the development of an insect, e.g., egg stage, larval stage, pupal stage, adult stage; in mites and ticks, each instar.

Stemma (pl., stemmata) The simple eye in holometabolous larvae. Also called lateral ocellus.

Sternite The plate or sclerite on the underside of a body segment.

Stigma A small colored area near the wingtip of dragonflies, bees, and various other clear-winged insects; also called the pterostigma.

Style A slender bristle arising at the apex of the antenna.

Stylet A needle-like object: applied to the various components of piercing mouthparts and also to a part of the sting of a bee or other hymenopteran.

Subcosta Usually the first of the longitudinal veins behind the front edge of the wing, although it is often missing or very faint: abbreviated to Sc.

Subimago Found only among the mayflies, the subimago or dun is the winged insect which emerges from the nymphal skin.

Superfamily A group of closely related families; superfamily names end in -oidea.

Suture A groove on the body surface which usually divides one plate or sclerite from the next: also the junction between the elytra of a beetle.

Tarsus (pl., tarsi) The insect's foot: primitively a single segment but consisting of several segments in most living insects.

Tegmen (pl., tegmina) The leathery forewing of a grasshopper or similar insect, such as a cockroach.

Tergite The primary plate or sclerite forming the dorsal surface of any body segment.

Tergum The dorsal surface of any body segment.

Thorax The middle of the three major divisions of the insect body.

Tibia (pl., tibiae) The fourth leg segment between the femur and the tarsus.

Trachea (pl., tracheae) One of the minute tubes which permeate the insect body and carry gases to and from the various organs. They open to the air at the spiracles.

Trochanter The second segment of the leg, between the coxa and femur: often very small and easily overlooked.

Tympanum The auditory membrane or eardrum of various insects.

Veins In insects, the rib-like tubes that strengthen the wings.

Venation The arrangement of veins in the wings of insects.

Vertex The top of the head, between and behind the eyes.

Vestigial Poorly developed, degenerate or atrophied, more fully functional in an earlier stage of development of the individual or species.

Wing pads The undeveloped wings of nymphs and naiads, which appear as two flat structures on each side.

Workers In termites, the sterile males and females that perform most of the work of the colony; they are pale, wingless, and usually lack compound eyes; in social Hymenoptera, females with undeveloped reproductive organs that perform the work of the colony.

INDEX

Acrididae, 122
Adephaga, 150
alderflies, 145, 146
Aleyrodidae, 144
ametabolous, 58, 59
Anisoptera, 115, 116
Anoplura, 136, 139, 181
antlions, 147
ants, 7, 8, 26, 52, 63, 70, 126, 127, 128, 149, 160, 161, 162, 205
Aphidae, 144
aphids, 137, 142, 143, 147, 161, 183, 185
Apidae, 163
Apocrita, 160, 161, 162
apodemes, 16
apolysis, 15
arachnids, 97, 206
Archaeognatha, 13, 19, 24, 30, 45, 58, 77, 109, 110, 182, 204, 213
archedictyon, 28, 111, 117, 125, 126
Arthropoda, 96, 98, 99, 100, 205, 211
Auchenorrhyncha, 142

Batesian mimicry, 76
beekeeping, 8, 10
bees, 5, 6, 7, 8, 9, 24, 26, 63, 65, 70, 71, 127, 128, 151, 158, 160, 161, 162, 187, 188, 210, 216, 218
beetle, 7, 8, 10, 46, 65, 153, 155, 157, 178, 208, 210, 219
Belostomatidae, 7, 141
Bible, 1
Blaberidae, 125
blastokinesis, 57
Blattaria, 95, 101, 117, 124, 185

Blattellidae, 125
Blatteria, 1
Blattidae, 125
Brachycera, 169
brain, 7, 19, 43, 45, 46, 47, 50, 52, 60, 61, 62, 207
bumblebee, 4
Buprestidae, 125, 157
butterflies, 16, 28, 37, 41, 58, 67, 69, 75, 76, 92, 93, 170, 171, 172, 177, 179, 187, 216
butterfly, 8, 170, 205, 206

caddisflies, 4, 170
Caelifera, 122
Calliphoridae, 169
Caloneurodea, 79
campaniform sensilla, 47, 49
Cantharidae, 157
Cantharidin, 7
Carabidae, 94, 150, 153
centipedes, 98, 99, 100
central nervous system, 43, 44, 45, 46, 47, 64, 70
Cerambycidae, 157
Cermbycidae, 125
Chelicerata, 99, 100
chemoreceptors, 48, 49
Chilopoda, 98, 99, 100
Chironomidae, 167
chitin, 12, 15, 184
chordotonal organs, 48, 67
chorion, 54, 55
Chrysomelidae, 157
Chrysopidae, 149
cibarium, 21, 24, 134, 141
cicadas, 1, 4, 7, 24, 69, 74, 142
Cicadellidae, 144
Cimicidae, 141

221

Coccoidea, 144, 183
cockroaches, 1, 2, 65, 68, 124, 125, 126, 127, 131, 185, 214
Coleoptera, 1, 7, 26, 28, 41, 65, 94, 96, 101, 125, 131, 149, 182, 187
Coleorrhyncha, 142
Collembola, 16, 55, 96, 98, 99, 100, 104, 105, 204
compound eyes, 49, 104, 108, 126, 170
Corixidae, 8
corpus allata, 60
Corydalidae, 146
Cretaceous, 1, 79, 128
crickets, 2, 4, 68, 69, 121, 122, 214
Crustacea, 99, 100, 108, 125
Culicidae, 94, 167
Curculionidae, 150, 153
cuticle, 12, 13, 14, 15, 16, 17, 25, 26, 30, 31, 33, 47, 48, 49, 51, 61, 62, 97, 104, 112, 161, 186, 206, 208, 209, 217
Cyclorrhapha, 169
Cydnidae, 125

damselflies, 58, 116, 182
Darwin, 5, 73, 74, 75, 95
DDT, 2, 4, 77
Dermaptera, 28, 57, 95, 131, 186
Devonian, 38, 77, 105, 109
diapause, 60
Diaphanopterodea, 78
Dicliptera, 78
Dictyoptera, 57, 117, 131
Diploglossata, 95
Diplopoda, 98, 99, 100
Diplura, 16, 58, 98, 99, 100, 104, 106, 204
Diptera, 2, 3, 4, 7, 41, 45, 69, 101, 142, 158, 166, 167, 182, 188, 210
direct muscles, 39
dobsonflies, 26, 145, 146

dorsal cl..., 57
dragonfl..., 58, 69, 111, 114, 1..., 149, 176, 182, 183, 2...
Dytiscid..., 150, 153

earwigs, ..., 32, 150
ecdyson...,
eclosion..., 61
economi..., evel, 84
economi..., ld, 84
ectopara..., 136, 165
egg, 29, ..., 57, 58, 60, 65, 67, 68..., 147, 187, 206, 207, 2..., 218
Elaterid...,
Embiidi..., 1, 119, 186
embryo, ..., 57, 60
Endopte..., 1
Ensifera,
Epheme..., 26, 27, 28, 53, 58, 111, 1..., 117, 181, 183
epicutic..., 15, 16, 209, 217
evolutio..., 8, 49, 58, 65, 73, 75, 77..., 98, 125, 127, 134, 1...
exoskele..., 1, 97, 98, 205, 217
eyes, 19, ..., 0, 104, 105, 108, 111, 1..., 117, 118, 121, 124, 1..., 132, 135, 136, 137, 1..., 151, 158, 164, 165, 1..., 210, 214, 220

fat body,
fireflies,
fleas, 2, ..., 95, 165, 166, 169, 1...
foregut, ..., 215, 216
Formici..., 26, 161, 162, 163
fruit flie...,
furcula,

gall, 6, 161, 210
Gerridae, 141
gills, 38, 58, 111, 112, 116, 118, 146, 210
Glosseyltrodea, 79
grasshopper, 21, 35, 36, 37, 48, 151, 219
Gryllacrididae, 122
Gryllidae, 122
Grylloblattodea, 26, 95, 101, 132, 181, 185
Gryllotalpidae, 122
gustation, 47

hardening, 15, 217
harvestmen, 97, 99, 100
hellgrammites, 146
hemimetabolous, 58, 59, 111, 118
Hemiptera, 1, 6, 7, 8, 125, 129, 135, 139, 142, 143, 181, 183, 188, 217
Hesperiidae, 172
Heteroptera, 141, 142
hindgut, 30, 212, 217
Histeridae, 150, 153
holometabolous, 59, 60, 79, 137, 145, 146, 147, 150, 162, 165, 212, 218
Homoptera, 196
honey, 5, 6, 8, 9, 10, 63, 66, 70, 71, 177, 210
honey bees, 162
honeydew, 7, 161
hormones, 31, 60, 61, 62
hornets, 162
horseshoe crabs, 97, 99, 100
Hydrometridae, 141
Hydrophilidae, 155
Hymenoptera, 4, 5, 7, 26, 55, 65, 66, 79, 96, 101, 126, 127, 160, 161, 162, 163, 182, 187, 220
hypopharynx, 19, 137, 206

Ichneumonidae, 163
indirect muscles, 39
innate behavior, 64
Insecta, 96, 98, 99, 100, 104, 206, 211
integrated pest management, 81
integument, 13, 14, 15, 17, 104, 105, 106, 131, 207, 208, 213, 218
Isopoda, 125
Isoptera, 8, 26, 57, 117, 125, 126, 161, 181, 185

Jurassic, 132, 162
juvenile hormones, 60

katatrepsis, 56, 57
kinesis, 64, 70

Labiata, 99, 100
lacewings, 26, 147, 149
Lampyridae, 157
lateral ocelli, 49
learned behavior, 64
Lepidoptera, 1, 5, 6, 7, 16, 26, 30, 62, 65, 96, 101, 160, 161, 164, 167, 170, 171, 172, 173, 174, 182, 187
lice, 1, 4, 26, 66, 134, 135, 136, 179, 183, 185
life table, 86
Linnaeus, 66, 91, 132
locomotion, 35, 167
locusts, 1, 2, 7, 8
Lucanidae, 155
Lycidae, 125, 157
Lygaeidae, 141

Mallophaga, 139
Malpighian tubules, 30, 33, 57, 117, 134, 147, 162, 212
mantids, 129, 131, 147, 158, 185, 214

Mantispidae, 129, 146, 147, 149
Mantodea, 101, 117, 129, 147, 185
Mantophasmatodea, 101, 124, 131
mayflies, 26, 53, 58, 69, 111, 112, 117, 118, 176, 219
Mecoptera, 164, 165, 166, 182, 186
Megaloptera, 26, 101, 145, 146, 147, 149, 182
Megasecoptera, 78, 111
Meloidae, 155
Membracidae, 144
metamorphosis, 19, 57, 58, 158, 204, 207, 208, 209, 211, 214, 216, 218
Micropterygidae, 171
midgut, 30, 33, 56, 57, 209, 210, 211
millipedes, 98, 99, 100
mimicry, 75, 76, 147
Miomoptera, 79
mites, 97, 99, 100, 205, 218
molting, 15, 60
Monarch, 75, 76, 171, 174
morphospecies, 92
mosquito, 3, 92
moth, 5, 6, 8, 22, 34, 67, 170, 172, 174, 180, 187, 205
moths, 1, 2, 16, 24, 26, 28, 41, 65, 67, 75, 76, 96, 170, 171, 172, 176, 177, 179, 187, 206, 209, 211
Mullerian mimicry, 76
Muscidae, 169
muscle, 15, 16, 31, 34, 35, 36, 42, 46, 61
Myrmeleontidae, 147, 149

natural selection, 73, 75, 77
Naucoridae, 141
Nematocera, 167
neopterous, 41, 117, 118
Nephrocytes, 31

Nepidae, 141
nerve impulse, 44, 46
nervous system, 41, 43, 44, 57, 108, 127, 208
neuron, 43, 44, 45, 46, 47
neurons, 43
Neuroptera, 26, 28, 101, 129, 146, 147, 149, 182
Neuropteroidea, 101, 145
Noctuidae, 172, 173, 174
Notonectidae, 141
Nymphalidae, 174

ocelli, 49, 66, 108, 111, 114, 118, 121, 124, 126, 129, 132, 134, 135, 137, 146, 147, 164, 165, 170, 214
Odonata, 26, 27, 28, 39, 41, 57, 58, 111, 114, 116, 117, 181, 182
oenocytes, 33
olfaction, 47
ommatidium, 49, 50
oothecum, 125
Orthoptera, 1, 4, 7, 57, 95, 101, 121, 122, 142, 186
ovaries, 53, 126
ovum, 55

Palaeodictyoptera, 78, 111
paleopterous, 41, 111, 117
Paleozoic, 12, 97, 145
Papilionidae, 174
paranotal lobes, 37, 78, 119
Passalidae, 151, 155
paurometabolous, 58, 59, 119, 125, 126, 132, 145
Pauropoda, 98, 99, 100
Pentatomidae, 7, 141
Permian, 38, 78, 79, 116, 119, 136, 143, 146, 149, 158, 165
Phasmida, 96, 101, 121, 186
pheromones, 62
Pieridae, 174

224

Plecoptera, 58, 112, 117, 181, 186
pleura, 17, 215
pollination, 4, 5, 65
polymorphism, 126
Polyphaga, 150
predator/prey models, 82
proctodeum, 30, 33, 56, 57
Protelytroptera, 79
prothoracic gland, 60
Protodonata, 78, 111
Protorthoptera, 79
Protura, 96, 98, 99, 100, 104, 106, 204
proventriculus, 30, 33
Psocodea, 1, 3, 101, 134, 181
psyllids, 135, 183, 185
pterothorax, 24, 114, 122, 125
Pterygota, 101, 216
pterythorax, 39
PTTH, 60
Pycnogonida, 97, 99, 100

Raphidioptera, 26, 146, 147
Reduviidae, 141
reflex arc, 46
reflexes, 46, 64
resilins, 13, 15, 41
retinaculum, 105
Rhipiphoridae, 158
roaches, 28, 79, 124, 125, 158

Saturniidae, 172, 174
sawflies, 160, 161
scale, 2, 6, 7, 16, 142, 143, 144, 177, 179, 183, 184, 185, 217
scarab, 65, 151, 155
Scarabeidae, 151
sclerites, 17
scolopophores, 48
scorpionflies, 164, 186
Scorpionida, 96, 99, 100
scorpions, 96, 97, 99, 100, 205
Sialidae, 146

silkworm, 6, 7, 62, 172
Silphidae, 153
silverfish, 1, 68, 109
Siphonaptera, 3, 26, 94, 95, 101, 165, 166, 182, 186
snakeflies, 26, 146
Sphecidae, 162, 163
Sphingidae, 67, 172, 173, 174
spiders, 97, 99, 100, 162, 186, 205, 206
spiracles, 30, 33, 47, 105, 216, 219
springtails, 99, 100, 105
Staphylinidae, 153
stemmata, 49, 51, 66, 212, 218
Sternorrhyncha, 143
stomodeum, 30, 57
stoneflies, 58, 96, 117, 118, 186
Strepsiptera, 101, 157, 158, 181
suboesophageal ganglion, 43, 45, 46, 62
sutures, 17, 18, 19, 136
Symphyla, 98, 99, 100
Symphyta, 160, 161, 162
Syrphidae, 169

Tabanidae, 169
Tachinidae, 169
tactile hairs, 47
tagmata, 97
tanning, 15
taxes, 64, 65, 70
Tenebrionidae, 151
Tenthredinidae, 161, 163
tentorium, 16, 21
Terebrantia, 139
termites, 8, 26, 28, 63, 124, 125, 126, 127, 128, 205, 214, 217, 218, 220
testes, 53
Tetrigidae, 122
Tettigoniidae, 122
thrips, 26, 136, 137, 139

Thysanoptera, 26, 101, 136, 139, 181, 185
Thysanura, 1, 13, 16, 19, 30, 36, 38, 58, 95, 109, 110, 182, 204
ticks, 97, 99, 100, 205, 218
tiger beetles, 187
Titanoptera, 79
trachea, 25, 30
Triassic, 78, 79, 162
trichogen, 16
Trichoptera, 4, 101, 170, 171, 173, 187
Trilobita, 99, 100
trophallaxis, 128
Tubulifera, 139

Upper Carboniferous, 25, 38, 78, 79, 106, 112, 122, 125

Vespidae, 161, 162, 163
Viceroy, 75, 76

waggle dance, 71, 72
walkingsticks, 121
wasps, 4, 5, 6, 26, 41, 63, 76, 127, 128, 147, 158, 160, 161, 162, 169, 176, 179, 187, 188, 210, 212
webspinners, 26, 119
weevils, 2, 150, 217

Xiphosura, 97, 99, 100

yellowjackets, 162
yolk, 54, 55, 56, 57

Zoraptera, 26, 124, 181, 185
Zygoptera, 115, 116

Made in the USA
Coppell, TX
02 June 2021